DEADly
OUTBREAKS

DEADLY OUTBREAKS

How Medical Detectives Save Lives Threatened by Killer Pandemics, Exotic Viruses, and Drug-Resistant Parasites

Alexandra M. Levitt, PhD

Foreword by
Donald R. Hopkins, MD, MPH

Skyhorse Publishing

Skyhorse Publishing books may be purchased in bulk at special discounts for
sales promotion, corporate gifts, fund-raising, or educational purposes. Special
editions can also be created to specifications. For details, contact the Special
Sales Department, Skyhorse Publishing, 307 West 36th Street,
11th Floor, New York, NY 10018 or info@skyhorsepublishing.com.

Skyhorse® and Skyhorse Publishing® are registered trademarks of
Skyhorse Publishing, Inc.®, a Delaware corporation.

Visit our website at www.skyhorsepublishing.com.

10 9 8 7 6 5 4 3 2 1

Library of Congress Cataloging-in-Publication Data is available on file.

Cover design by Liz Driesbach

Print ISBN: 978-1-63450-266-5
Ebook ISBN: 978-1-5107-0526-5

Printed in the United States of America

Dedication

To Patrick McConnon, Kate McConnon, and Scott Fischer—
and to medical detectives at state, local, and federal public
health agencies throughout the nation.

Contents

Foreword

In the latter part of the twentieth century, some believed antibiotics and vaccines had led to the conquest of most infectious diseases, and we could then turn our full attention to non-infectious, chronic conditions. Later, alarming experiences with HIV/AIDS, SARS, a virulent strain of influenza, multiple-drug resistant tuberculosis and other newly-recognized maladies corrected that misimpression. We now recognize that the struggle between humans, microbes and other aspects of our environment is long and ever-changing, but never ending.

Alexandra Levitt's book, *Deadly Outbreaks*, is a fascinating account of seven frightening public health mysteries encountered in the United States between 1976 and 2007. Five of the episodes resulted from infectious diseases, ranging from an old foe, malaria, to a more recently recognized one, hantavirus; others were due to an autoimmune condition and drug overdoses in a cluster of infants. These are gripping, suspenseful stories that are exceptionally well-written and highly instructive. Public health practitioners and students will benefit from the hard-won victories of epidemiologists described here, and in my opinion, one can hardly hope to learn important lessons for the future in a more enjoyable way. At the end of each episode, the author reviews and describes explicit lessons illustrated by the events and the reactions to them. Beyond the interesting science and methodology of epidemiology *per se*, the reader experiences vicariously the personal and professional mindsets of epidemiologists as they deal with substantial medical, bureaucratic, political and legal challenges, sometimes simultaneously, often while being well aware of the high-stakes consequences of their findings. A frequent theme is to remind us all of how complex processes associated with modernization can create new pathways for spread of disease or injury. Another theme concerns the value and importance of communication between practitioners of different specialties, such as physician-epidemiologists and veterinarians, for example.

This volume is in the finest proud tradition of Paul de Kruif's *Microbe Hunters* (1926), which described the discovery of various microbes, and Berton Rouché's *Eleven Blue Men* (1953), which described outbreaks of different diseases. Like the latter, *Deadly Outbreaks* describes the learn-by-doing approach to the practice of epidemiology by "disease detectives" at the U.S. Centers for Disease Control and Prevention's Epidemic Intelligence Service (EIS) and at state and local health agencies. It is a timely reminder that the battle with infectious diseases and other outbreaks is an on-going struggle, as epidemiologists in each age are challenged by newly-emerged pathogens, newly-realized pathways for old pathogens, and other unpredictable surprises. And for all the welcome and powerful new tools of computers, faster communications and more sophisticated statistical techniques, we see how much still depends on inquiring minds and dogged determination.

> Donald R. Hopkins, MD, MPH
> Vice President, Health Programs
> The Carter Center

Author's Note, 2013

I wrote *Deadly Outbreaks* at the request of my friend and colleague Patrick J. McConnon, the former Executive Director of the Council of State and Territorial Epidemiologists (CSTE), to showcase field epidemiology as an exciting and rewarding career choice for students who are interested in science and math. I worked with Pat for many years at the Centers for Disease Control and Prevention (CDC), where he served as Associate Director for Emerging Diseases Program Development of the National Center for Infectious Diseases. In that capacity, Pat not only strengthened medical detection by designing innovative training programs, disease surveillance programs, and fellowships, but also handled operations and helped coordinate complex outbreak responses, including the CDC's responses to the 1993 hantavirus outbreak in the southwestern United States, the 1997 Rift Valley Fever outbreak in Kenya, and the 2001 anthrax incidents in Washington, DC. I dedicate *Deadly Outbreaks* to Pat, with affection and gratitude, as well as to Pat's wife, Kate, and my husband, Scott. I also dedicate this book to medical detectives who work at public health departments throughout the country.

As a government employee, I would also like to make the following disclaimer:

The opinions and views expressed in this book are my own and do not reflect the views of the Centers for Disease Control and Prevention, the Department of Health and Human Services, or the U.S. Government.

Author's Note, 2014

Disease detectives learn something new from nearly every investigation of an infectious disease outbreak. In writing this introduction to the paperback edition of *Deadly Outbreaks,* I thought it might be useful to review investigations conducted since the book was published—and to consider what these experiences have taught us. Here are my lessons learned:

One: Prepare for the Unexpected. The diversity of infectious disease outbreaks in the U.S. between September 2013 and December 2014 confirms that—when it comes to infectious microbes—we must always be ready for the unexpected. This fall, for example, a strain of enterovirus—a common type of virus that typically causes mild cold-like symptoms—unexpectedly caused severe respiratory disease in children (especially children with asthma) that required hospitalization and intensive care.[1] Several unusual foodborne outbreaks occurred as well. Last winter, more than 630 people in twenty-nine states fell ill with salmonellosis after eating meals prepared from a contaminated frozen chicken product purchased at supermarkets across the country. The causative strains of bacteria were resistant to several common antibiotics. Another multi-state outbreak of gastrointestinal illness, which affected more than three hundred people, was traced to imported batches of fresh cilantro contaminated with *Cyclospora*, an intestinal parasite endemic to the tropical and sub-tropical regions of the world.

Something else unexpected: The *New York Times* recently reported on the surprising legal aftermath of a nationwide investigation of fungal

---

[1] Over the course of the fall and winter, physicians reported an unusual number of cases of children hospitalized with muscle weakness or paralysis. As I write, an investigation of these cases—which do not appear to be associated with enterovirus infection—is underway.

meningitis that was winding down when *Deadly Outbreaks* was published.[2] More than a year earlier, medical detectives had determined
that a medical steroid product used to treat lower back pain and sciatica had been contaminated with a common mold called *Exerholium
rostratum*. Injected into patients' spines and joints, the mold caused
spinal infections, meningitis, and strokes, leading to sixty-four deaths.
On December 17, 2014, The *New York Times* stated that two executives of the Massachusetts compounding pharmacy that prepared
the contaminated product have been charged with twenty-five acts of
second-degree murder in seven states.

Two: We Are All in It Together. Over the past year, the United
States was directly affected by two epidemics that are causing
devastation in other parts of the world. Last spring, two Saudi healthcare workers on vacation in the U.S.—one in Indiana and one in
Florida—fell ill with a dangerous, SARS-like disease caused by the
Middle East Respiratory Syndrome coronavirus (MERS-CoV). Both
were hospitalized, and both survived. In October, a traveler from
Liberia visiting friends and family in Texas was hospitalized with Ebola
hemorrhagic fever, three days after being sent home from the emergency room because of an initial misdiagnosis. The traveler, Thomas
Eric Duncan, died a little over a week later. Two nurses who cared for
him contracted Ebola fever; fortunately, both survived. Since then, one
other person in the U.S. has been diagnosed with Ebola (an American
doctor returning from Guinea), and five Americans who contracted
Ebola while working on medical response teams in Liberia and Sierra
Leone have been evacuated to the U.S. for treatment.

The outbreaks of MERS and Ebola fever illustrate that—especially
when it comes to infectious diseases—we live in a global village where
everyone is potentially exposed to everyone else. Helping other nations
control outbreaks at their source is therefore a matter of self-interest
as well as a humanitarian imperative. Infectious microbes, which have
always traveled between countries and continents—in people, animals,

---

[2] Jess Bidgood and Sabrina Tavernise. *Pharmacy Executives Face Murder Charges in
Meningitis Deaths. New York Times*, December 17, 2014.

insects, and objects—can spread faster than ever before in today's world of air travel and global trade. Therefore all of us are at risk, no matter where we live.

Three: We Are Fortunate to Live in a Country with a Strong Public Health System. My final observation is that each outbreak that occurred over the past year, whether at home or abroad, confirms the major take-home lesson of *Deadly Outbreaks*—that we are fortunate to live in a country with a public health system that we can take for granted as being there when we need it. The Ebola outbreak in West Africa, in particular, demonstrates in stark and painful terms what can happen when a society lacks a routine way to detect unusual diseases, isolate and treat infectious patients, and prevent disease from spreading in homes, hospitals, and the community.

With these ideas in mind, I dedicate the paperback edition of *Deadly Outbreaks* to the brave and resourceful medical detectives and laboratory scientists who work behind the scenes to keep us safe.

Alexandra M. Levitt, PhD
December 30, 2014

Author's Note, 2020

As I write, the world is experiencing an outbreak of Coronavirus 2019 (COVID-19) disease, a new respiratory disease that we are learning about as it spreads from country to country. This disease causes fever, cough, and shortness of breath, and sometimes progresses to pneumonia that can be life-threatening.

COVID-19 is caused by a type of virus called a coronavirus, which has a circular shape and surface spikes that look like halos or coronas. Until the twenty-first century, coronaviruses were associated with the sniffles and aches of the common cold. In 2003 and 2012, however, two previously unknown coronaviruses "jumped" from animals to humans, causing outbreaks of severe (and sometimes fatal) respiratory diseases. Severe Acute Respiratory Syndrome (SARS) spread overnight by airplane from Hong Kong to Canada, Vietnam, and Singapore, and Middle East Associated Respiratory Syndrome (MERS) spread through the Arabian Peninsula, with travel-related cases reported in twenty-seven countries.

Like SARS and MERS, COVID-19 arose from an animal reservoir. The viruses that cause COVID-19 and SARS came from animals sold in live animal markets (pangolins and civets) in large cities in China, while the virus that causes MERS came from camels used as pack animals. All three viruses are maintained in bats, with civets, pangolins, and camels serving as intermediate hosts. The animal origin of these viruses underscores the need for a "One Health" approach to human and animal medicine.

As with SARS in 2003, the virus that causes COVID-19 emerged in a densely populated urban area.

Government authorities in the city of Wuhan implemented highly aggressive quarantine measures that slowed the spread of COVID-19 but could not contain it. To date, cases of COVID-19 have been reported in more than eighty countries, on every continent but Antarctica, with documented community spread in Singapore, Hong Kong, South Korea, Italy, Japan, and the United States. As compared with SARS seventeen years ago, the wider global spread of COVID-19 likely reflects a greater degree of interconnectedness between China and other countries, as well as a worldwide increase in international travel, trade, and tourism. More than ever before, we live in a global village where everyone is connected to everyone else.

Another difference from 2003 involves changes in modern communications that affect us in both positive and negative ways. On the plus side, social media and cable TV are rapidly disseminating information that helps individuals and families protect themselves from disease. Based on what we know so far, people at higher risk of severe illness and death—*older people and people with underlying health conditions such as diabetes and lung, heart, or kidney disease*—are urged to practice everyday preventive actions recommended for everyone (washing hands, covering coughs and sneezes, cleaning frequently touched surfaces) and to avoid crowds and people who are ill until the time of danger has passed. Although people who experience severe disease will likely require hospitalization, for most people COVID-19 symptoms are mild and self-limiting, and homecare with rest and fluids is sufficient.

Rapid dissemination of accurate public health information is very important. Unfortunately, some social media platforms are spreading panic and misinformation, forwarding conspiracy theories, and even setting us against one another by scapegoating groups of people, such as Asian citizens and travelers. This is dangerous and divisive at a time that requires collective action based on the understanding that—when it comes to outbreak responses—we are all in it together.

Fortunately, disease detectives around the world are working together to develop, share, and utilize advanced scientific tools. Scientists at the Chinese Centers for Disease Control rapidly isolated the virus that causes COVID-19, sequenced its genome, and posted the viral sequence on GenBank, which is curated by the US National Institutes of Health. Scientists in many countries then used this information to develop diagnostic tests based on amplification of unique viral sequence fragments. (This process is described in Chapter 7, which harks back to a time when the amplification technique was new.) Moreover, clinicians in the earliest affected countries are sharing what they have learned about optimizing care for people with severe disease.

Collaborative efforts are also underway to develop vaccines and antiviral drugs against COVID-19, with public/private partnerships, manufacturing platforms, and funding sources already in place that aim to shorten the time it takes (at least a year) to confirm that a vaccine or drug is both safe and effective. Five years ago, during the Ebola outbreak in West Africa, an international group of responders set a valuable precedent when they fought the outbreak and developed countermeasures at the same time. Today, a vaccine tested during the waning days of the outbreak in West Africa is saving lives endangered by an Ebola outbreak in the Democratic Republic of the Congo.

Meanwhile, in the absence of vaccines and drugs against COVID-19, local response efforts in the United States are benefitting from "nonpharmaceutical interventions" (NPIs) developed as part of preparedness planning for pandemic influenza. NPIs include personal protective measures like the everyday preventive actions described above, as well as social distancing measures for communities, such as cancelation of mass gatherings and school closures. During the H1N1 influenza pandemic in 2009, we learned more about the practical side of implementing these measures in US communities with different resources, needs, and attitudes. The United States is therefore better prepared to respond to COVID-19 at the community level.

A non-health aspect of the outbreak of COVID-19 is its devastating impact on the US and global economies, with disrupted supply chains, cancelled air flights, stock market vacillations, and decreased patronage at small businesses and restaurants. Public concern is at a fever pitch, and people are hunkering down in their homes, even in communities where cases of COVID-19 have not been reported. Only at a future time, after the outbreak has passed, will we be able to take stock of our losses and judge whether the disruption to social and economic life was a necessary evil to stop a dangerous epidemic or an over-reaction to a scary but manageable situation.

As a government employee, I would like to make the following disclaimer:

The opinions and views expressed in this book are my own and do not reflect the views of the Centers for Disease Control and Prevention, the Department of Health and Human Services, or the US government.

Alexandra M. Levitt, PhD
March 8, 2020

Information about COVID-19

Centers for Disease Control and Prevention (CDC):

- About COVID-19: https://www.cdc.gov/coronavirus/2019-ncov/about/index.html
- Updates on the outbreak of COVID-19: https://www.cdc.gov/coronavirus/2019-ncov/summary.html
- Information for Travelers: https://www.cdc.gov/coronavirus/2019-ncov/travelers/index.html
- Information for People at Risk for Serious Illness from COVID-19: https://www.cdc.gov/coronavirus/2019-ncov/specific-groups/high-risk-complications.html

World Health Organization: https://www.who.int/emergencies/diseases/novel-coronavirus-2019

Introduction

An epidemic of infectious disease is a dreaded event that evokes fears of danger, mortality, and the unknown. This is true even today, when we have known for more than a century that microbes rather than evil spirits, rotting food, or bad air cause disease. Unlike most natural disasters, whose terrors are confined to a particular time and place, an epidemic proceeds slowly and invisibly, in a silent, unpredictable, and inexorable way, lasting for weeks or months or even longer. The unsettling idea of an active, hidden, and malevolent force is reflected in the word we use to describe a newly emerged epidemic—an *outbreak*—which suggests that something dangerous and sinister has broken out of its restraints, like a monster escaped from a dungeon or a madman from a prison for the criminally insane.

No wonder that an outbreak can make us feel unprotected and helpless, especially when the basic facts—where the outbreak started, what microbe caused it, and how it is spread—are unknown. You cannot easily run away from an invisible enemy, or barricade your family and friends in an attic or basement until the epidemic is over. And if the cause of the epidemic is a new or drug-resistant microbe, even the experts may be unable to tell us (at least at first) how to protect ourselves and our loved ones.

Thus, microbes remain a disturbing prospect to most people and a formidable adversary to scientists and doctors. Despite healthcare advances such as vaccines and antibiotics, modern medicine has not been able to "conquer" infectious microbes, because of their amazing ability to change, adapt, evolve, and spread to new places. As the Nobel Prize winning microbiologist Joshua Lederberg liked to say. "Pitted against microbial genes, we have mainly our wits."[1] Our microbial adversaries not only overcome the obstacles we put in their way (e.g., by developing drug resistance) but also take advantage of modern technologies that allow them to spread in new ways and flourish in new niches. Microbes can live in modern ventilation systems, travel to new continents by airplane, and contaminate centrally-processed food products shipped to stores and restaurants in different locations. Microbes also benefit from increasing urbanization, mis-use of antibiotics, and increased opportunities to "jump" from animals to humans as more people use rain forests and other wilderness areas for settlement, agriculture, recreation, or tourism. Finally, as we learned during the anthrax incidents in 2001, human beings can deliberately spread pathogenic microbes as a weapon of warfare or terror.

Terrorism aside, some of the greatest dangers from infectious disease we face today are from microbes that develop drug resistance—like tuberculosis or staph or strep bacteria—and from new microbes that emerge unexpectedly from animal reservoirs, like the viruses that cause AIDS, SARS, and pandemic influenza. Because these diseases are new, we do not always have tools to treat them or prevent their spread. I remember the shock and disbelief I felt in 1983 when a young man I knew, a lively, charming graduate student in his late twenties, died within a few weeks of falling ill with a rare fungal pneumonia, now known to be an opportunistic infection of AIDS. His doctors' helplessness recalled the days before antibiotics and vaccination, when deaths from infectious disease were common. I imagined what it must have been like in 1923, when my own grandfather, a doctor in New York City, died of bacterial meningitis contracted from a patient. There was no vaccine and no treatment. This is what we potentially face—even today, in the twenty-first century—with each outbreak of a new or newly drug-resistant disease.

What type of person is ready and able to pit his or her wits against the endless inventiveness of infectious microbes? The goal of this book is to answer this question by describing the scientific adventures of a special group of people—known formally as *field epidemiologists* and informally as *medical detectives* —who investigate outbreaks and figure out how to stop them, working in close partnership with public health laboratory scientists. These curious, determined, and intrepid individuals tend to be an unusual hybrid of "people-person" and scientific nerd. Like homicide detectives, they witness human suffering and elicit information from people in acute distress as part of their basic duties. Like physicians and nurses, they put themselves at risk while helping patients who may carry contagious diseases. Unlike a policemen or healthcare provider, however, they typically investigate many cases at once as part of a single mission, trying to figure out what all the cases have in common. In each case, their methodology involves the collection of medical and public health data that is carefully recorded, stripped of its human sorrows, and analyzed in the most wonkish of ways, using graphs, tables, and (above all) statistics. Public health statistics—"people with the tears wiped away"[2] —are the stock-in-trade of medical detectives, who have unshakable faith in their power to uncover the disease patterns behind the grief, loss, and pain caused by an epidemic. In many cases, medical detectives use their statistical findings not only to control disease but also to find ways to prevent future outbreaks once the immediate emergency is over.

The successes of these brave individuals often go unremarked and unsung. Most are employed by state and local health departments, and their work typically occurs behind the scenes, unknown to the general public. Most of the time, when they prevent disease spread, illness or death, no one is aware of it. As a result, the importance of what they do may be overlooked, and their jobs may be endangered by budget cuts, especially at times of recession and belt-tightening. I hope that *Deadly Outbreaks* helps to attract the next generation of epidemiologists by highlighting the indispensable role of medical detectives in maintaining the public health system that protects the nation's health.

1

Dead Crows Falling
from the Sky

Medical detective Annie Fine was reluctant to venture out to Queens for a routine check on a few encephalitis patients at a neighborhood hospital in Flushing. The patients suffered from an inflammation of the brain that is typically triggered by some type of infection, and it was her task to make sure that the cases were not linked. She felt sure the long weekend of work would result only in a hodgepodge of unrelated cases—and a big waste of time. However, she was the epidemiologist "on call" for the New York City health department, and her friend and boss Marcelle Layton—Assistant Commissioner of the NYC Bureau of Communicable Diseases—had asked her to come along. To Fine's surprise, it turned out to be one of the most fascinating episodes in her career, equal in interest to the investigation of the anthrax incidents of 2001, still two years in the future.

Something Strange

Though Fine did not realize it at the time, there actually would be two investigations—one about humans and one about birds—that moved on parallel tracks for a few hectic weeks and then converged into a surprising (and satisfactory) conclusion. The human track began on Monday, August 23, 1999, when Dr. Deborah S. Asnis, chief of infectious diseases at Flushing Hospital in Queens, called the NYC health department to report two cases of encephalitis in elderly patients, which included fever, headache, and mental confusion.

Both patients had been previously active and healthy. One patient had a fairly unusual condition, flaccid paralysis (extreme muscle weakness) of the type that used to be associated with polio but is now more often a sign of botulism poisoning. He required a mechanical ventilator to breathe. Based on the totality of the symptoms, it looked like viral encephalitis. But if his illness were botulism (a potentially fatal bacterial disease), antitoxin should be administered right away, without waiting for test results. By Thursday, the second encephalitis patient was on a ventilator too.

New York City's Communicable Diseases Program actively encourages doctors to call when they observe something unusual. Although the Assistant Commisioner's staff was fairly small, her policy was to have a doctor respond to each call. This time, Layton, an internist trained in infectious diseases, took the call herself. She enjoyed keeping her hand in by speaking with local clinicians like Asnis, who (as one of the only infectious diseases specialist for adults in Queens) saw a lot of interesting cases. Asnis said the patient's muscle weakness was not of the descending type, beginning in the upper body and gradually spreading downward, as expected with botulism.

Because the NYC public health laboratory did not test for viruses, Layton asked Asnis to send blood and cerebrospinal fluid samples to the virology laboratory at the New York State (NYS) Department of Health, which had initiated a program for testing specimens from people with undiagnosed encephalitis, in association with the Centers for Disease Control and Prevention (CDC). Layton also sent a staff member to Queens to review the patient's chart and confirm that the symptoms suggested viral encephalitis rather than botulism. Asnis had noticed that several spinal taps had been ordered that summer to test for encephalitis or meningitis in elderly patients at Flushing Hospital. This stood out to her, as typically only two or three are ordered per year, most often when meningitis is suspected in a child or young adult.

Four days later, on Friday, August 27, Asnis called Layton to report a third case of encephalitis with muscle weakness in an elderly person at Flushing Hospital. During the call, she was interrupted by a neurologist colleague who had just returned from the New York

Hospital Medical Center of Queens, where he had seen a fourth case of encephalitis, diagnosed as Guillain-Barré syndrome, an autoimmune condition sometimes triggered by infection. The coincidence was definitely strange! Layton spoke with foodborne disease experts at the CDC, who agreed that Asnis' cases did not sound like botulism.

Fewer than ten cases of encephalitis are reported in New York City in an average year. Yet within the span of a week four patients from a small corner of Queens had been diagnosed and put on ventilators. Something was not right. Though fieldwork is not a regular part of an assistant commissioner's job, Layton cancelled her weekend plans and called in Annie Fine.

The Investigators

Both Marci Layton and Annie Fine are primary-care doctors who came to New York City on two-year assignments as "EIS officers," trainees in the CDC Epidemic Intelligence Service, the national training program for medical detectives. Layton, who grew up in the suburbs of Baltimore, became actively interested in science and medicine in her early teens. Inspired by the social ideals of the 1960s, she volunteered at hospital emergency rooms, local nursing homes, and summer programs for the mentally and physically disabled. As she grew from teenager to young adult, an ongoing succession of volunteer projects and short-term jobs opened her mind to the diversity of other lives—rich and poor, sick and well, American and foreign—-and to the diversity of medical practice. Her experience in rural areas in the United States reinforced her love of the outdoors, while her overseas experiences in Asia revealed the boundless possibilities of international travel. Both provided an intimate view of the huge impact of infectious diseases in poor communities.

As a medical student, Layton's volunteer adventures included a summer at a rural clinic in West Virginia, where she roomed with a woman called Granny Parsons who had never before met a Jew or ventured more than a few miles from her "holler." Layton also worked on the Navajo Reservation in the Four Corners area of the American southwest, where one of her mentors was Bruce Tempest, the Indian Health Service physician who (a few years later) reported the first cases of a previously unknown respiratory disease (see Chapter 7).

After her first long-term post-residency job—an unsatisfying stint in an understaffed family-care clinic in Providence, Rhode Island—Layton completed an infectious disease fellowship at Yale and decided to enter public health. When she joined EIS in 1992, Thomas Frieden, the outgoing EIS fellow assigned to NYC—who later became the director of the CDC—recommended New York as a place where an epidemiologist can work on a variety of diseases without leaving home. Layton thought of her EIS assignment as another short-term venture and did not expect to like New York, which she associated with the fancy shops and skyscrapers of midtown Manhattan. But over time she came to enjoy the unending challenges of working at the NYC health department, where she was hired as Assistant Commissioner upon completing her EIS fellowship. Today she is the "go-to" person for anyone seeking the big-city perspective on infectious disease issues in public health.

Annie Fine also came to public health after working as a primary-care doctor. Fine—who is currently settled in Brooklyn and is the mother of twins—had lived in many different East Coast cities as a child and teenager. After completing a residency in pediatrics in San Francisco, she planned to remain in the Bay Area to work in medically underserved areas of Richmond and Oakland. However, she was unhappy practicing medicine under the managed-care system, where she felt like a factory worker on a human conveyor belt, with fifteen minutes to "fix" each child. Seeking a more effective way to use her training, she moved back east and entered the EIS program.

Fine loved New York City and loved working with Layton. Moreover, she soon began dating another doctor, a dermatologist raised in New Jersey, who admired and supported her decision to continue working in public health. But there was only one job open at Communicable Diseases when her EIS training ended, and Fine was not sure she wanted it. The Department of Health was looking for a person to serve as liaison to the Office of Emergency Management

(OEM), which was developing a city-wide bioterrorism response plan and organizing emergency response exercises.

When Fine completed her EIS training—three years before the terrorist attacks of September 11, 2001—New York was already deeply engaged in terrorism preparedness activities. The 1993 World Trade Center bombing confirmed NYC's attractiveness as a target of terrorism, and the 1995 nerve gas attack on the Tokyo subway underscored its vulnerability as a big city. The Port Authority of New York and New Jersey took steps to increase security procedures at the World Trade Center, and Mayor Rudolph Giuliani, who took office in 1994, made terrorism preparedness a priority. In 1996, Giuliani initiated plans to build the ill-fated Emergency Command Center and bunker at 7 World Trade Center, across the street from the twin towers. He also shifted responsibility for emergency preparedness from the NYC police department to the newly established OEM, whose first leader was Jerry Hauer, a former emergency management administrator for the State of Indiana. A high-powered man with a forceful personality, Hauer was more successful at competing for resources with New York's Finest than any of the OEM directors who have so far succeeded him.

Although Fine was not happy about switching from disease investigations to bioterrorism preparedness, she didn't want to leave her professional connections or life in New York City. However, despite its crazy hours, the liaison job proved unexpectedly absorbing and interesting. Fine worked with police, firefighters, and hazmat personnel on planning drills that simulated such scenarios as plague attacks, truck bombs, and hoaxes. She also made presentations to community and professional groups. In August 1999, her main focus was a huge, million-dollar, live-action drill called "CitySafe," scheduled to take place in a quiet Bronx neighborhood in September, with volunteers acting as victims stricken by airborne anthrax spores. The drill was a major production—federally-funded, the first of its kind—with high-level participation from the U.S. Army. It was even rumored that President Clinton might attend.

CitySafe was only two weeks away when Layton asked Fine to accompany her to Flushing Hospital.

The Field Investigation

On Saturday morning, August 28, Fine and Layton drove to Flushing Hospital in northern Queens. As Layton got out of the passenger side of the car, near the curb, she stepped over a large, dead black bird—an image that would recur in the days ahead. They went straight to the intensive-care unit, where the chief resident awaited them, standing in for Asnis, who was out of town taking care of a family emergency. He described the encephalitis cases, reviewing each medical chart in detail.

It was immediately evident how unusual and how similar the cases were. The patients were all older people, all previously healthy and active. Although encephalitis is often accompanied by inflammation of the meninges, the membranes that surround the brain and spinal cord, these patients did not exhibit typical meningitis symptoms such as severe headache and a stiff neck. The patients experienced fever, stomach pains, nausea, and diarrhea, followed by mental confusion, muscle weakness, and breathing difficulties, necessitating ventilatory support. One patient came to the emergency room thinking he was having a heart attack because of the breathing difficulties and pain. In each instance the illness developed slowly, over several days to a week. In contrast, the symptoms of botulism poisoning usually occur within eighteen to thirty-six hours of eating a contaminated food.

The laboratory clues were also remarkably consistent. In each case, analysis of the cerebrospinal fluid indicated the presence of white blood cells and a higher-than-normal level of protein—two signs of infection. Under the microscope, the white blood cells were identified as primary lymphocytes, which is a sign of viral rather than bacterial infection.

Once the chart review was complete, Layton and Fine interviewed the patients' families. Strangely enough, not long after they began, a shouting elderly patient was wheeled in on a stretcher, angry and combative. He, too, had fever and encephalitis: a possible fifth case of the mysterious disease.

Layton and Fine were unable to talk directly with the patients at Flushing Hospital, because three were on ventilators and the newly admitted patient was delirious. Fortunately, the team was able to rely instead on information from each patient's worried family visitors. Eager for any type of lead, Fine and Layton began by asking what the patients had been doing the week before they fell ill, hoping to find an activity they might have in common. Did they attend parties, visit relatives, attend church, go shopping, go to a movie, or go to a restaurant? What did they eat and what medications did they take? What were their hobbies? It quickly emerged that the patients and their families were not acquainted with each other. The only common factor was that they all lived within a densely populated area of about four square miles in Queens.

As they spoke with the patients' families, Layton and Fine were struck once again by similarities among the four cases. The patients were elderly, but not at all debilitated. They were active people who spent considerable time outdoors. One was tanned from long hours of gardening in his backyard. Another took long walks every day. A third had been exiled to the porch by her son, who didn't want her to smoke indoors. The fourth patient liked to sit in his front yard, watching and kibitzing as a swimming pool was built next door.

Afterwards, Layton—a considerate boss—asked Fine to drop her off at the New York Hospital Medical Center of Queens and go home to Brooklyn while Layton visited the last patient. It was now late in the day, visiting hours were over, and the New York Hospital Medical Center was fairly empty. This patient, like the others, was on a ventilator and unable to speak. Layton quickly reviewed his chart, which told a similar story, with a similar progression of symptoms and laboratory findings from the analysis of the cerebrospinal fluid.

Troubled, but hopeful, Layton walked to the Flushing subway station. Passing the Queens Botanical Garden, she walked in on impulse to collect her thoughts. It was not yet dark, and the minaret of a mosque was visible above the trees. She sat on a bench in the Rose Garden and took stock of what she needed to do next. Suddenly hungry, having missed both lunch and dinner, she entered the first eating place she passed after leaving the Garden, a Vietnamese restaurant, and tried something new: Vietnamese iced coffee, served hot but poured over ice, strong and sweetened with condensed milk.

Layton returned to her office and called Fine at home to tell her about the patient at the New York Hospital Medical Center. Layton and Fine were now more certain that something strange was going on, and that there was no time to lose in figuring out what it might be. They would contact the CDC in the morning and also check in with other NYC hospitals to find out if there were any similar cases. By mid-morning Monday, Layton's staff had identified four more cases of encephalitis with muscle weakness—in Queens and the South Bronx—bringing the total to nine.

Taking Stock

Fine and Layton reviewed local data to determine whether the encephalitis cases were as unusual, statistically speaking, as they seemed to be. One hundred to 120 sporadic cases of viral meningitis are reported in NYC each year, with the number of cases increasing in July as the temperature rises, and then going back down. However, the small number of reported NYC cases of viral encephalitis without meningitis—only seven or eight per year—makes it difficult to draw conclusions about seasonality, although an upswing in August did not seem atypical. (Reporting of both diseases depends on voluntary reporting by healthcare providers, and both diseases are generally believed to be underreported.)

Next, Fine and Layton reviewed medical textbooks for information about the symptoms, signs, and differential diagnoses of viral encephalitis. They confirmed that flaccid paralysis is not a common manifestation of most types of viral encephalitis and that the diverse (and ubiquitous) members of the Enterovirus genus are the most common viral cause of encephalitis. However, all of the Queens patients were elderly, and enteroviral infection occurs mostly in babies and children, possibly because older people acquire immunity to enteroviruses through years of mild infections. (In contrast, "hot babies" seen in emergency rooms in July and August often have enteroviral infections.) Moreover, enteroviruses typically cause meningoencephalitis (combined encephalitis and meningitis), rather than encephalitis alone. Nevertheless, it was possible that the Flushing cases involved an uncommon enterovirus with unusual clinical manifestations.

Layton and Fine were aware of other, more exotic causes of viral encephalitis reported overseas, such as Nipah virus (a paramyxovirus related to measles and mumps), which had been identified the previous spring in Malaysia and Singapore. Moreover, the leading cause of viral encephalitis in Asia—the mosquito-borne Japanese encephalitis virus—had spread into Australia less than a year before. Japanese encephalitis is one of several encephalitic viruses that are zoonotic (transmitted from animals to humans) and arboviral (transmitted by arthropods, such as mosquitoes and ticks). Like enteroviruses, arboviruses most often cause meningoencephalitis rather than encephalitis alone. Although arboviral diseases have not been a problem in New York since the yellow fever outbreaks of the 1800s, Layton and Fine knew that unusual illnesses from other parts of the world sometimes turn up in New York City.

Later that day, Fine's fiancé found a tiny paragraph in *Harrison's Principles of Internal Medicine* on a mosquito-borne viral disease called St. Louis encephalitis (SLE), which occurs most often in elderly people and can cause muscle weakness and mental confusion. Looking back, Fine says, "David was the first to get it wrong."

St. Louis encephalitis in Queens? The idea seemed very far-fetched. SLE had never been reported in New York City and is so rare upstate that it was not on the list of diseases monitored by the New York State Arthropod-Borne Disease Program.[1] The virus that causes SLE is a member of the *Flavivirus* genus (family *Flaviviridae*) carried by birds; its overseas "cousins" include the Japanese, Kunjin, Powassan, and West Nile encephalitis viruses, as well as two hemorrhagic fever viruses known worldwide: dengue and yellow fever. About thirty sporadic cases of SLE are reported in the United States each year, with occasional outbreaks in hot, wet places like the Mississippi Valley and the Gulf Coast.

Nevertheless, Layton and Fine agreed that a mosquito-borne disease would be consistent with the patients' histories of outdoor activities. Coincidentally it had also been good weather for mosquitoes, with a mild winter, wet spring, and unbearably hot, dry July (the hottest on record). In fact there had been two unusual incidents of mosquito-borne malaria, one involving two eleven-year-olds at a Boy Scout camp in Long Island. Moreover, the last time NYC had experienced similar weather conditions, during the summer of 1993, Layton's office had investigated three cases of malaria in adults in Queens. However, there were no surveillance data to confirm an upswing in the local mosquito population, because NYC's pest control program had been sharply reduced in the early 1990s. Although some funding was restored in 1997, it was mostly allotted to rat control.

Consultation

Layton called the CDC's twenty-four-hour emergency hotline number on Sunday morning. Not certain which division to contact—since the causative agent of the encephalitis cases was unknown—she asked to speak to Ali Khan, a CDC colleague she had worked with in the past. But the CDC operator did not put her through to Khan, who was currently in charge of bioterrorism surveillance. Instead, she was connected with an expert on arboviruses and then (on a second try) with an expert on enteroviruses. Neither one was interested, because the Flushing cases did not fit the expected profile for either arboviral or enteroviral infection. Still worried, Layton sent an e-mail to Khan, and copied the others. Khan responded by arranging a Monday morning conference call that included scientists with expertise on enteroviruses, arboviruses, bioterror agents, and "special pathogens" (a catch-all category covering rare and unknown microbes).

On the phone that morning, the CDC experts dismissed the idea of a mosquito-borne outbreak in New York City. They recommended that a neurologist review the cases, ignoring the fact that a neurologist had examined Asnis' patients before she called Layton. They asked Layton to send blood and cerebrospinal fluid samples to the CDC, as well as to New York State.

A second call did not go well either, although Khan offered to send an EIS officer from the CDC's Bioterrorism Surveillance

Branch to assist in the investigation. By this time, Layton's staff had turned up several additional suspected cases—the first trickle of what would soon become a flood—by phoning specialty departments at each city hospital (adult and pediatric emergency medicine, intensive care, infectious disease, neurology, and infection control practitioners) and by issuing a "broadcast fax" (a useful mechanism in the days before e-mail came into common use) to physicians and nurses throughout the city. After the phone call ended, Layton and Fine heard their CDC colleagues talking on the still-open line: It's not an outbreak. . . . There would be meningitis as well as encephalitis. . . . There have been no outbreaks of arboviral diseases in New York City for more than a hundred years. . . .

Still mystified, and wanting to assess the possibility of a mosquito-borne disease, Layton asked Veruni Kulasekara, a medical entomologist at the American Museum of Natural History, to help evaluate the mosquito situation in Queens. Kulasekara, a chic Sri Lankan married to an American rock star (the base guitarist of the Violent Femmes), had met with Layton earlier that summer to discuss the potential risk of local malaria transmission, given the hot, wet weather. Accompanied by an EIS officer and a public health veterinarian, Kulasekara set off Tuesday morning to examine the homes of Asnis' patients (as well as other newly identified patients with unexplained encephalitis) for mosquito breeding sites. The patients lived in the adjacent neighborhoods of College Point and Whitestone, on the northern coast of Queens. College Point is located just above Flushing (the site of Shea and Arthur Ashe stadiums), close to LaGuardia Airport, and separated from the South Bronx by a narrow stretch of the East River; Whitestone is just west of College Point. Both neighborhoods contain block after block of modest-but-comfy one- and two-family homes with front lawns and small backyards, bordered by concrete sidewalks with openings for maple trees and flower beds.

That evening, Layton went to Fine's apartment in Park Slope, Brooklyn, to await the return of the EIS officer on Kulasekara's team, a chronic disease specialist named Denis Nash who was filling in for a colleague from the Bureau of Communicable Diseases. (New to NYC, Nash had never been to Queens and knew little about mosquitoes.[2] However, his EIS training had taught him how to follow the disease trail wherever it might lead.) While they waited, they ordered in Thai food and discussed the invitation list for Fine's upcoming wedding in Brooklyn's Prospect Park.

Arriving just in time for a late-evening dinner, Nash reported that the NYC investigators had no difficulty finding mosquito breeding sites in the College Point area. There were piles of tires with water in the doughnut holes, and backyard swimming pools with just enough stagnant water to create attractive habitats for mosquitoes. They found an open plastic barrel in the backyard of the sickest patient—the avid gardener whom Layton and Fine had visited at Flushing Hospital—who had been collecting rainwater to use on his plants. In a scoop of rain-barrel water Kulasekara showed them mosquito larvae of the *Culex* genus, a type of mosquito that can transmit arboviruses such as SLE virus. One member of this genus—*Culex pipiens*, the northern, or common, house mosquito—is frequently seen in urban areas.

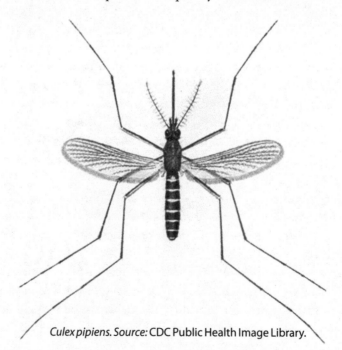

Culex pipiens. Source: CDC Public Health Image Library.

After discussing Nash's conclusion—that a mosquito-borne disease was a distinct possibility—they turned to the practical problem: how to triage and investigate the growing number of suspected cases, which rose to more than sixty by week's end. Lacking a diagnosis, they would have to rely on "syndromic surveillance,"[3] which is case-finding based on a characteristic group of signs and symptoms rather than on laboratory testing. They devised a case definition on the basis of four clinical criteria: fever, altered mental status, muscle weakness, and laboratory test results consistent with a viral cause. A patient whose illness fit all four criteria was designated as a definite case; a patient who fit three, as a probable case; two, as a possible case; and one, as an unlikely case.

At the office the next day, Layton assigned several field investigators—including Denis Nash (borrowed from the Bureau of HIV/AIDS); the EIS officer assigned to Communicable Diseases; and a third EIS officer sent, as promised, by Khan—to interview the families of patients with definite and probable cases, review medical records, assess clinical findings, and collect demographic data (age, sex, and occupation). They also collected blood and cerebrospinal fluid samples that were packaged by the NYC health laboratory and sent to the CDC and to the NYS encephalitis laboratory. The EIS officer from the CDC took charge of the paper-based line listing of suspected cases and transferred it into an Excel spreadsheet, along with the demographic, clinical, and laboratory clues.

There was more bad news as the week went on. The avid gardener died on Tuesday evening, and the woman who smoked on her porch died a few days later. Fine called Hauer to give him a heads-up that a deadly encephalitis agent might be carried by mosquitoes. If so, it could be a serious problem for CitySafe—the big, live-action bioterrorism preparedness drill—which was supposed to take place outdoors.

The Diagnosis

On Thursday, two days before Labor Day weekend, preliminary test results arrived from the New York State virology laboratory. To their amazement, blood samples from two patients tested positive for

SLE—just as Fine's fiancé had suggested! Was this really true? The state laboratory had used an indirect test called the immunofluorescent antibody (IFA) assay, which detects antibodies to particular viruses or bacteria. Although IFA can give false positives (when antibodies specific for one microbe cross-react with antibodies specific for another), all of the clues (medical, entomologic, and laboratory) fit together fairly well: encephalitis in older adults with muscle weakness (a rare SLE symptom, but mentioned in *Harrison's*); *Culex* mosquitoes; and an IFA positive for SLE. It was certainly possible that a NYC mosquito had taken a blood meal from an SLE-infected bird that migrated in from a southern state. But it was still very odd to have an SLE outbreak in NYC, where there had never been a case before—especially when no cases of SLE had been reported elsewhere in the country.

There is no drug treatment for SLE, so medical care is supportive, aimed at mitigating the most serious symptoms (brain inflammation and breathing difficulties) until the patient recovers on his or her own. From a public health perspective, the main response is mosquito control—not a small or inexpensive undertaking, especially with no pre-existing program in place. Because this diagnosis was so surprising and unexpected, the NYC health department decided to wait for confirmation from the CDC, promised for the next day, before taking action. Duane Gubler, head of the CDC Fort Collins laboratory, was using a different blood test, the virus-specific, enzyme-linked immunoabsorbant assay (ELISA), which gives fewer false positives than IFA. The ELISA test panel would include North American arboviruses from three viral genera: *Alphavirus* (eastern, western, and Venezuelan equine encephalitis viruses), *Orthobunyavirus* (La Crosse and California encephalitis viruses), and *Flavivirus* (Murray Valley and Rocio viruses, as well as SLE). All of these viruses are endemic to the United States, but most are uncommon in the northeast.

Sick with suspense, Fine waited for the official diagnosis from the CDC. When the call came confirming SLE, Fine raced down the hall, feeling like Paul Revere rousing the troops: It's positive! It's positive! Now that they knew what was going on they could do something about it. Layton spoke with Hauer, who put the resources of the OEM and Emergency Command Center at the health department's disposal. A few days later, Hauer canceled CitySafe.

The Public Health Response

The diagnosis came on the Friday before Labor Day weekend, a time for traditional outdoor activities such as backyard barbeques, block parties, and concerts in city parks. Mayor Giuliani quickly held a press conference to get the word out about the outbreak and the measures the city was taking to rid the city of mosquitoes. Fine rode in the health department's official black sedan with psychiatrist Neal Cohen, the NYC Health Commissioner, to meet Giuliani and Hauer in College Point.

It was an uncomfortable ride. The commissioner was nervous, and the driver was unsure of the directions. Fine was on the car phone with Layton, who was back in her office, talking on two lines at once, participating in a conference call with the CDC, whose experts were recommending a pesticide called malathion. Fine was concerned about the decision. She had been in California in 1994 when malathion was sprayed on medfly-infested orchards, causing environmental damage. She relayed the CDC's advice to the commissioner, explaining about the pesticide issues in California in case there were questions from reporters.

In College Point, while they waited for Giuliani to arrive, a neighborhood resident came up to Fine and asked her a strange question: Had she heard about the dead birds in Queens? Lots of crows and robins? Did she think there was a connection between the dead birds and the sick people? Fine said she didn't know anything about the birds. SLE is carried by birds, which seemed like an obvious connection, but Fine was pretty sure that SLE doesn't kill birds, which are its natural host and reservoir.

Fine briefed Mayor Giuliani, who relayed the public health information to the reporters and the crowd of local residents. He described the three confirmed cases of SLE, expressed sadness at the death of the gardener, and asked everyone to cooperate in the city's effort to "wipe out the mosquito population." Flanked by Hauer and Cohen, he announced that several neighborhoods in northern Queens and nearby areas of the South Bronx would be sprayed in the evening and around dawn, when *Culex pipiens* mosquitoes are most active. He suggested that city residents, especially the elderly, stay indoors or wear pants and long sleeves when going out. He demonstrated the use of insect repellent—splashing it on his face and arms—and asked that people rid their yards of standing water from pools, tires, or other containers surfaces.

Later, Fine returned to the office where she and Layton and their colleagues planned out what they would need to do in the coming week and how to make the best use of OEM resources. Their immediate concern was to send the latest set of patient specimens to Fort Collins without delay. Because FedEx would not deliver on the holiday weekend, OEM chartered a plane to deliver the specimens to a private home in Fort Collins, where the father-in-law of a CDC scientist would be waiting to sign for them.

Fine spent the rest of the day at a newly activated OEM hotline at the Emergency Command Center at 7 World Trade Center, which had opened the previous June. The Command Center was a modern, high-tech facility with corporate décor—a far cry from the shabby, institutional-yellow offices of the NYC health department. Fine sat in a comfortable, ergonomically correct office chair and answered call after call from local doctors and worried parents. The call she remembers best was from a woman in Queens who asked her the same question as the woman at the press conference: What was going on with the birds? The woman quoted from a neighborhood paper that described "droves of crows falling from the sky" in Queens and Nassau County, Long Island, in the Bronx and even further north, up into Westchester. Was there a connection with the mystery disease in College Point? Fine said she didn't know, but she would find out.

That task, however, would have to wait until tomorrow; it had been a very long day. When Layton and Fine returned home to Park Slope, by subway, very late, Layton dropped Fine off on her stoop and hugged her good-bye.

The Birds

Fine called Roger Nasci, a CDC vector-borne disease expert, the day after the press conference to follow up about the birds. What did he think was going on? It seemed unlikely that the bird and human outbreaks were simply a coincidence, but there were a few things that did not fit. First, SLE virus is carried by birds, but it doesn't make them ill. Second, the birds were dying over a fairly large area—Queens, the Bronx, and Long Island—but human encephalitis cases had been reported only in Queens.

Nasci is a mild-mannered, unpretentious scientist, described by one of his colleagues as salt of the earth. He was not disparaging or dismissive when he reiterated that SLE would not sicken the birds that are its natural reservoir. But he admitted that he had no idea what type of illness was killing the birds. He said he was coming to Queens to measure the density and distribution of *Culex pipiens* and any other mosquito species that might be transmitting SLE. He would also help New York City use surveillance data to determine whether the malathion spraying was effective. When he later arrived at LaGuardia Airport, a police escort was waiting for him, courtesy of OEM, with lights flashing and sirens blaring.

Fine also called the New York State Arthropod-Borne Disease Program to ask whether its staff would test a dead crow dropped off at an OEM mobile command unit in Queens. They referred her to Ward Stone at the Wildlife Pathology Unit at the state's Department of Environmental Conservation in Albany. A wildlife enthusiast with a mad-scientist swirl of white hair, Stone's duties included performing necropsies (animal autopsies) and sending tissues to veterinary laboratories for testing. For the NYC health department, he mostly conducted toxicology tests on dead pigeons possibly poisoned in city parks.

Stone accepted the NYC shipment but did not mention that he had already received lots of dead birds—four hundred or more through the month of August—and was sufficiently alarmed to alert his counterparts in New Jersey and Connecticut. The dead birds came from zoo officials in Queens and the Bronx, as well as from private citizens, including a highway cleaning crew in Long Island. Tracey McNamara, the head pathologist at the Bronx Zoo—a research veterinarian with an intense and animated way of speaking—had called him several times. Two weeks earlier, on August 19, Stone told her he had identified several different causes of illness but "no common thread"[4] that would explain why the birds were dying.

Tracey McNamara had much reason for concern. On Wednesday, August 25—three days before Layton and Fine began their investigation in Flushing—an owl and an eagle living in an outdoor cage at the Bronx Zoo had died of unknown causes. More zoo birds died over Labor Day weekend after exhibiting strange behaviors, such as a wobbly gait, tremors, and abnormal head postures. The dead birds included Chilean flamingoes, a ring-necked pheasant, and a Guanay cormorant that swam in circles for several hours before it died. McNamara was especially worried because one of her assistants stuck himself with a syringe needle while trying to medicate a jerking flamingo.

By the time Fine contacted Stone, McNamara had already ruled out two major killers of U.S. poultry (Newcastle disease and avian influenza) on the basis of routine blood tests. She had dissected several birds and found signs of encephalitis (i.e., bleeding from the brain) and damage to the heart muscle. This led her to consider eastern equine encephalitis, a mosquito-borne encephalitis virus. However, equine encephalitis is highly dangerous to emus (a large, flightless bird, related to the ostrich), and the Bronx Zoo emus were fine. Besides, encephalitis suggested a connection to the human cases in Queens. McNamara called the CDC on the Thursday after Labor Day, described the necropsy data, and asked for help in taking care of her assistant.

The head of the CDC epidemiology unit in Fort Collins asked her to send a serum sample from the employee and a plasma sample from the sick flamingo to help evaluate the assistant's risk of contracting disease. (Fortunately, the employee did not develop any symptoms.) However, he was not used to working with zoo officials and (like his colleagues) did not take her concerns about the human-bird connection seriously. He told her the same thing Nasci told Fine: bird die-offs have many possible causes, and the two outbreaks were probably a coincidence. He went on to say that his laboratory was extremely busy with the human SLE cases in New York City and did not have time or resources to investigate illness in birds. He referred her to the National Veterinary Services Laboratory (NVSL) in Ames, Iowa, which is the major U.S. Department of Agriculture laboratory concerned with farm animals. This was the same laboratory recommended to Ward Stone by the veterinary laboratory at Cornell University, a back-up facility to the New York State virology laboratory, which (like the CDC) does not have diagnostic reagents or responsibility for testing diseases in birds and animals.

As time went on and more zoo birds died (including ducks, laughing gulls, and magpies), McNamara—unable to interest the CDC, despite calling every day for a week—set in motion an extensive veterinary investigation that expanded to include another state laboratory (the Connecticut Agricultural Experiment Station), the federal agricultural laboratory (NVSL), and two federal laboratories concerned with wildlife and bioterror agents. The wildlife laboratory was the National Wildlife Health Center (NWHC) in Madison, Wisconsin, which is operated by the U.S. Geological Survey, Department of the Interior, and the bioterrorism laboratory was part of the U.S. Armed Forces Institute for Infectious Diseases (USAMRIID) in Fort Detrick, Maryland. Although USAMRIID generally does not take requests from civilians, a USAMRIID veterinary pathologist eventually agreed to accept Bronx Zoo bird specimens as a special favor to McNamara.

Layton and Fine would not hear about these activities for another two weeks, by which time the veterinary investigation had circled back to Fort Collins. By Thursday, September 23, three of the veterinary laboratories—NVSL, NWHC, and the Connecticut Agricultural Experiment Station—had isolated viruses they could not identify and had called the CDC for help.

The Mosquitoes

Rising to its first major challenge involving a public health emergency, the NYC Office of Emergency Management had a mosquito-control

program underway within twenty-four hours. This was a significant achievement because OEM had to start from scratch, without
city pest-control experts or mosquito surveillance data to guide its
actions. After consulting with specialists from Suffolk County, Long
Island, and the CDC, OEM purchased huge amounts of malathion
and borrowed a Suffolk County airplane to survey Queens for potential mosquito breeding sites. Helicopters were rented, loaded, and
readied for immediate use. OEM advised NYC residents to stay
indoors with windows closed and air conditioners turned off for two
hours after spraying.

Mosquito-control measures were front-page news in all local
papers. The health department tripled its staff at the hotline at 7
World Trade Center and provided health department personnel
with their first cell phones. Throughout Labor Day weekend, the
health department and OEM sent mobile command units to talk
with people in affected neighborhoods; distribute insect repellent at
police and fire stations; and disseminate press releases, fact sheets, and
flyers. Because the risk to any given person was small, no events—
except CitySafe—were canceled. Nevertheless, health officials urged
residents of Queens (especially the elderly) to avoid outdoor celebrations and encouraged people attending Central Park concerts,
the U.S. Open at Arthur Ashe Stadium, or baseball games at Shea
Stadium to wear skin-covering clothes and use insect repellent (some
provided courtesy of the Mets).

Lacking historical data on mosquito habitats—and with new
surveillance efforts just underway—the health department and OEM
used the locations of confirmed human SLE cases to decide where
spraying would do the most good. They anxiously awaited each set
of diagnostic reports from the arbovirus laboratory at the CDC,
which was providing results on individual patients using the ELISA.
(This was a departure from the CDC laboratory's usual role as a
public health reference facility, but the NYS virology laboratory did
not have experience with the ELISA test for SLE.) Each afternoon,
Layton, Fine, Hauer, and and health department colleagues squeezed
into a third floor conference room with an open speakerphone to

hear John Roehrig—their main CDC contact—read off daily case results, one by one. (Roehrig, who had not worked with the NYC health department before, at first confused Fine's and Layton's voices on the phone and referred to them as "the sisters." In person, he would have seen that Layton and Fine are dissimilar in look and manner, though both are petite brunettes. Layton is wavy-haired, quietly observant, and diplomatic in her persistence, while Fine is curly-haired, outgoing, and direct.)

Everyone's ears pricked up when Roehrig announced a positive case in Sunset Park, Brooklyn—the first case outside of Queens and the South Bronx. Cell phones flicked open all over the room, and people began to get up and leave. Layton, sitting at the head of the table, was soon the only one left. "Got to call you back," she said and hurried to her tiny office, where she found Hauer and Cohen and some of their staff crammed into the space around her desk, staring at her paper wall-map with its color-coded push pins. The Brooklyn case was twelve miles away from the cluster of cases in Queens and the Bronx, which might mean that the outbreak was spreading.

Layton sent an EIS officer to interview the Brooklyn patient, a Central Asian immigrant who spoke Russian and Farsi. As luck would have it, the EIS officer assigned to Communicable Diseases was an Iranian-American who spoke fluent Farsi and had no trouble conversing with the patient and his family. He confirmed that the patient had not left Brooklyn for weeks, which meant that he had been infected in Brooklyn and not in Queens or the Bronx.

On the following day, September 9, a third patient at Flushing Hospital died from viral encephalitis. Cohen informed Layton that Giuliani was holding a second press conference to announce that spraying would be expanded to cover Brooklyn and Manhattan. The commissioner wanted her to attend and bring her famous wall-map.

It was a rainy, cold day, and Layton's EIS officer wrapped the fraying wall-map—pushpins and all—in plastic. She stood by Cohen's side as he showed Giuliani the map and the updated list of definite, probable, and possible cases. "These numbers don't add up," Giuliani barked. The EIS officer was shocked, unsure what he meant.

Layton then sat down beside Giuliani and reviewed the numbers, case by case. The mayor backed down and apologized. After that, the reputation of the health department rose at City Hall, because the mayor reportedly told his staff that the health department was OK: "Their numbers always add up."

Over the next ten days, cases of unexplained encephalitis were confirmed throughout the city, and pesticide was sprayed in all five boroughs. In Manhattan, where skyscrapers interfered with the aerial approach, ground-spraying trucks were used in place of helicopters.

An Unresolved Issue

The next several weeks were a blur of activity at the NYC Bureau of Communicable Diseases. Layton's staff continued to conduct active case-finding, investigate suspected cases, and track laboratory results, which was a bookkeeping nightmare because the CDC and the NYC health department used different sets of identifying numbers. They also continued to prepare daily case updates for Giuliani, checking each patient, each day, by phone, to get a status report. If they learned that a patient had died, they arranged for autopsy tissue samples to be sent to the NYS and CDC laboratories. Later on, they conducted retrospective surveillance to determine whether the outbreak had been underway before Asnis noticed the cluster of cases in Flushing. Retrospective surveillance involved tracking hospital discharge diagnosis codes for paralysis, encephalitis, and neurologic conditions. Just one retrospective case was found: in a woman hospitalized in Queens during the first week in August.

Layton's staff also continued to assist the mayor's office with press releases and TV appearances. During the run-up to another press conference, Fine—in full last-minute wedding-planning mode—squeezed in a thirty-minute fitting for her vintage wedding gown. Layton accompanied her so they could continue refining the mayor's talking points. The people in the dress shop were fascinated by the conversation and by Fine's noisy work paraphernalia—three beepers and a cell phone— which she peeled off her belt before trying on the dress.

Meanwhile, Layton and Fine were also grappling with an unresolved issue—in addition to the mystery of the birds—related to the laboratory testing of the suspected SLE cases. It seemed fairly easy to spot a highly probable case of SLE, because the constellation of symptoms characteristic of the initial patients (fever, mental confusion, muscle weakness) was quite distinctive. (A case was only classified as "definite" after it was laboratory-confirmed.) Nevertheless, many probable cases came back as "indeterminate" or even "negative." Something did not seem right. As Roehrig rattled off one day's results, Fine wondered if there could be something wrong with the test. Roehrig, though also puzzled, treated it as a joke: "Would you quit saying that!!"

As it turned out, public health experts at the NYS virology laboratory were also unhappy with the SLE results. On Monday, September 13, Fine was invited to participate in the first day of a teleconference on encephalitis cases of unknown origin, associated with the CDC Unexplained Deaths Project. In view of recent events, the participants—from state laboratories in California, Minnesota, New York, and Tennessee—were eager to hear about the SLE cases in New York City. The New York State contingent told Fine they had been unable to confirm their SLE-positive IFA results using the polymerase chain reaction (PCR) technique. This didn't make sense because PCR is more direct and specific than an antibody-based blood test, involving amplification of a tiny piece of viral genetic material (a gene fragment) characteristic of a given virus or virus strain. As far as they knew, CDC had not confirmed the IFA and ELISA results either. In fact, CDC had not reported any SLE-positive results obtained with PCR, viral isolation studies, or the plaque-reduction neutralization assay, a quantitative test that measures the ability of the antibodies in a patient's serum to decrease viral activity. (Although plaque-reduction is also a "serologic" test, based on detection of antibodies, is less prone than IFA and ELISA to false positives caused by cross-reactions with antibodies from other flaviviruses.) It was also surprising that SLE virus had not been isolated from autopsy samples, though this might be because the level of virus in SLE-infected human brains is usually very low. (Eventually, the CDC did try to confirm the ELISA results with plaque-reduction assays, but the results were negative for SLE.)

On the last day of the teleconference (not attended by Fine or Layton), the New York State scientists decided to send autopsy tissues to W. Ian Lipkin, a molecular biologist at the University of California at Irvine. For the encephalitis workgroup, involving Lipkin would solve two problems at the same time. It would be a good test of the PCR-based methods Lipkin was developing for identifying encephalitis viruses, and it might shed light on the reasons for the discrepancies in SLE testing.

The Veterinary Investigation

Layton and Fine still had no information on what was killing the birds. The bird illness did not seem like a major issue at the time, because they accepted what the experts told them—that SLE did not kill birds and that the avian and human illnesses were unrelated. They waited to hear more and meanwhile encouraged further testing of dead birds that came to the attention of the health department. On Friday, September 10, Fine asked a vertebrate ecologist at the CDC for his take on the situation. (She had learned nothing further from Ward Stone, although by this time, urged on by McNamara, he had sent Bronx Zoo and Department of Environmental Conservation bird samples to the federal agricultural and wildlife laboratories [NVSL and NWHC]). The ecologist still thought the bird and human outbreaks were a coincidence, but he decided to visit Queens himself to get a look firsthand. He was planning to take blood samples from live birds, rather than dead ones, in an effort to identify host species carrying SLE, as well as figure out the cause of the mysterious avian illness. He did not see how SLE could be making the birds sick.

Before his bird survey was well under way, however, a major clue emerged from another source: the Connecticut Agricultural Experiment Station, which had revived its mosquito-control program in 1996 to cope with an outbreak of eastern equine encephalitis. Theodore Andreadis, an entomologist at the Station, had found a dead crow near a puddle full of mosquitoes on a golf course in Greenwich. When he necropsied the bird he found clear evidence of encephalitis. (Like Layton and Fine, he was unaware of similar findings by

McNamara in zoo birds.) He was now trying to isolate a virus from the crow's brain and also from the mosquitoes.

As it happened, Andreadis was not the only one trying to isolate a virus from a bird that had died from encephalitis. By mid-September, scientists from both NVSL and NWHC, unable to detect a known animal pathogen in Stone's and McNamara's bird specimens, had begun their own viral isolation studies. They injected bird tissues into embryonated chicken eggs and looked for viruses in the eggs' allantoic fluid. Under the electron microscope, both laboratories saw the same thing, a virus of the same genus as SLE: a flavivirus. Andreadis' laboratory also isolated flaviviruses from the golf-course crow and from the near by mosquitoes. All three laboratories contacted Duane Gubler in Fort Collins.

Andreadis and the NWHC scientists wondered if this might be a new, bird-killing variant of SLE, or perhaps an entirely new virus never seen before. The NVSL scientists made an additional suggestion: Could it be a human pathogen not previously known to infect birds? (The opposite situation, where a known animal pathogen is found to cause illness in humans, is not uncommon. Examples include monkeypox, Rift Valley Fever, and avian influenza A[H5N1]viruses.) The upshot was that NVSL sent its flavivirus—isolated from the brain of a Bronx Zoo flamingo—to the CDC.

By this time, having spoken multiple times with McNamara, Andreadis, and colleagues at NVSL, NWHC, and USAMRIID (who had also isolated a non-SLE flavivirus)—and having viewed the plaque-reduction results—Gubler was convinced that the initial diagnosis of SLE had been premature. The disease agent was certainly a flavivirus but *which* flavivirus was unclear.

Gubler is a friendly, outgoing man—a genial combination of down-home, small-town American and world-renowned public health authority. Raised in a tiny Utah town, he married a local girl who expected to be a rancher's wife but ended up a world traveler when her husband developed an interest in tropical diseases. They had lived in Hawaii, India, Indonesia, and Puerto Rico, where Gubler investigated mosquito-borne diseases such as filariasis and dengue hemorrhagic fever. For the past ten past years, he had headed the CDC Fort Collins laboratory, which studies diseases carried by insects and animals and provides outbreak assistance to state and local health departments throughout the United States.

Since Labor Day, Gubler had followed the standard Fort Collins investigative protocol for arboviral outbreaks, deploying a field ento-mologist to collect mosquito specimens and a vertebrate ecologist to identify host animals (in this case, birds) for further testing. Now, as he waited for the NVSL virus to arrive, he and Roehrig began re-testing the human blood samples from New York City, using an expanded ELISA panel that included Old World flaviviruses from Africa (e.g., West Nile virus), Asia (Japanese and Powassan viruses), and Australia (Kunjin virus). As they talked over the new developments, they agreed that brushing off McNamara had been a major mistake.

Resolution

Everything came together quickly after that. On Thursday, September 23, Gubler called Layton to let her know that the CDC would announce new findings during a conference call the next morn-ing. Layton and Fine assumed that meant a new diagnosis—probably a different flavivirus—and they speculated about what it could be. The unusual hush-hush quality of Gubler's message inspired wild thoughts to go through their minds, including the possibility of bioterrorism.

Layton spent all night working in her office and flew to Maryland early the next morning to fulfill a prior engagement as a panelist on a USAMRIID videoconference on the public health response to bio-terrorism. It was uncomfortable to answer questions about the NYC outbreak when she didn't know what the punch line was going to be. Afterwards, a fellow panelist—C.J. Peters of the CDC Special Pathogens Laboratory—speculated that the virus might be a close relative of SLE, such as West Nile virus (WNV) or Kunjin virus.

After her panel ended, Layton found a small private office in the studio and dialed into the CDC conference call—as did Fine and the rest of the team in NYC—and heard McNamara's story for the first time. They were both fascinated and a bit surprised to learn about the veterinary investigations that had taken place without their knowledge. McNamara described the bird deaths at the Bronx Zoo, her necropsy data, and her repeated efforts to get federal authorities to take notice. She made an especially interesting observation that explained how a virus whose natural host is birds could kill so many of them. The birds that died belonged to New World species like the Chilean flamingo and the Guanay cormorant. If the mysterious flavivirus were an Old World pathogen, unknown in the Americas, local birds would have no natural immunity to it. In that case, the introduction of an alien virus into the local bird population would be the veterinary equivalent of the catastrophic introduction of smallpox into Native American villages in the 16th century. Not so surprising, then, that wobbly, disoriented birds would "fall out of the sky in droves."

Gubler spoke next. He said that the CDC had confirmed the detection of West Nile virus in birds and suspected that the same pathogen—West Nile virus or a "West-Nile-like" virus—might be causing the human outbreak as well. But they would not be sure until his laboratory had re-tested blood and cerebrospinal fluid samples from several of NYC's highly probable but SLE-negative or indeterminate cases, using the expanded ELISA panel.

Gubler said that his laboratory was also comparing a large (1278 base-pair) viral gene fragment amplified from NYC birds, mosquitoes, and human autopsy samples, to further confirm that the NYC-area bird and human outbreaks were due to the same arbovirus. As Layton learned later, Gubler's laboratory had already begun sequencing the complete eleven thousand base-pair genome of the flamingo virus isolated by NVSL, starting with the gene for a structural ("envelope" or "coat") protein called the E-glycoprotein. Meanwhile, he had sent a one hundred base-pair E-glycoprotein gene sequence to two flaviviral experts—J. S. MacKenzie at the University of Queensland, Australia, and Vincent Deubel at the Pasteur Institute in Paris—who would soon confirm that it was an almost perfect match to the equivalent sequence from a West Nile virus strain isolated from an Israeli goose in 1998. The goose sequence had not yet been published, possibly because a scientist had taken the goose from Israel to France illegally (i.e., without declaring it to customs officials). The E-glycoprotein sequence from the NYC flamingo isolate was also a close match to viruses isolated from humans and mosquitoes during a Romanian WNV outbreak in 1996.

Gubler concluded the phone conference by explaining that the CDC planned to announce the findings in birds on Saturday, but wait until Sunday to announce the new human diagnosis, to have sufficient time to generate new ELISA test results on hundreds of suspected human cases.

At last! Layton's and Fine's remaining questions were answered. The outbreak of human illness was caused not by SLE virus but by West Nile virus, which explained the bird deaths and resolved the discrepancies between the clinical data and the laboratory findings. The exotic West Nile virus had multiplied in crows and other susceptible birds, and the mosquitoes that fed on the blood of infected birds had passed the virus on to humans.

Layton and Fine were relieved to realize that NYC officials and health-care workers were already doing what was necessary to treat patients and prevent disease spread. WNV is an Old World cousin to SLE—found in Africa, the Middle East, India, France, and the northern Mediterranean area—that usually causes milder symptoms in humans and has a slightly lower fatality rate (3–10 percent instead of 3–15 percent). Supportive care for WNV encephalitis and SLE is the same. Moreover, WNV is transmitted by the same *Culex* mosquitoes that transmit SLE (night-biters that breed in polluted water) but is also thought to be carried by *Aedes vexans* mosquitoes (day-time-biters that breed in rivers). Efforts to decrease the numbers of NYC mosquitoes that could carry SLE virus or WNV were already well under way (and by late September, temperatures were beginning to fall, along with mosquito counts).

However, the day's surprises were not yet over. Arriving back in New York, Layton received a confusing call from Lipkin, the scientist at the University of California, Irvine—whom she had not met or heard of before—who invited her to join him in a press conference about the new diagnosis. He spoke very fast and she had difficulty following him, but she thought he said that the cause of the human outbreak was a mixture of West Nile and Kunjin viruses, and that the viruses could have been intentionally released. She was not sure she understood him correctly.

She phoned her office, the NYS health department, and then the CDC. Her NYS colleagues explained about sending Lipkin the autopsy tissues for PCR testing. While the veterinary laboratories had focused on the bird samples, Lipkin had studied the ones from humans. He had isolated RNA from frozen brain tissues and devised a PCR technique, as Gubler had done, to amplify a gene fragment that varies in a characteristic way among different flaviviruses. He then compared the amplified sequences to sequences in GenBank—the National Institutes of Health databank that contains all publicly available DNA and RNA sequences—and found two near-matches: West Nile virus and Kunjin virus, a closely related virus endemic to Australia. He was hoping to announce his results as soon as possible. (Lipkin didn't know about the unpublished sequence from the Israeli goose virus, which was the reason that Gubler and his colleagues were able to classify the NYC virus as West Nile.)

Layton declined to participate in Lipkin's press conference. Although she did not care whether Lipkin or Gubler received credit, she wanted to make sure New Yorkers received the same information about WNV from New York City, New York State, and the CDC.

That night, she was awakened at 1 AM by an official from the health department press office who called to tell her that a *New York Times* reporter had requested confirmation of a statement from Lipkin that would run in the morning paper. Layton phoned Gubler in Fort Collins, where it was 11 PM, to give him a heads-up. The next morning, Lipkin's account of a mixture of two viruses appeared in the early edition of The *New York Times*, and a new version ran in the afternoon edition with a different headline and quotes from both Lipkin and Gubler.[5]

An official NYC-NYS-CDC press conference took place on Sunday morning, as scheduled, with several big guns in attendance, including Mayor Giuliani, Commissioner Cohen, and NYS Health Commissioner

Antonia Novello. Layton and Fine were on hand to answer questions. They had spent all day discussing McNamara's data and reviewing the CDC's figures for the new case count, which had more than doubled as old cases (from NYC and Nassau and Westchester counties) were re-classified as positive. The day before, with the human-bird connection confirmed, Layton had called an emergency meeting of the entire Communicable Diseases Program staff (except for Fine, who was ironing out the final details for her fast-approaching wedding). They needed to set up a surveillance system for dead birds. Before, they had used the location of human cases (a "human surveillance system") to decide where to target mosquito-control efforts. Now they would use dead crows (a "dead bird surveillance system") to identify neighborhoods where humans were most at risk. Dead bird jokes were the order of the day.

The Aftermath

Two weeks later, Fine and her fiancé were married outdoors, as planned, at dusk, in Prospect Park. The days had begun to get colder, and there were fewer mosquitoes, though the day itself was unseasonably warm. They were reassured by the CDC's estimate (incorrect, according to a later study)[6] that only one out of one thousand mosquitoes carried West Nile virus. They joked about supplying the guests with "goody bags" containing insect repellent.

As it turned out, they need not have worried, because the outbreak was already over, with no new cases of human illness identified after the third week in September. The final tally was sixty-two people with laboratory-confirmed West Nile disease, out of a total of 710 suspected cases in New York City and 229 suspected cases outside of New York City. Thus, only about 6 percent of the suspected cases turned out to be caused by West Nile virus. Of the sixty-two persons with confirmed cases, all but three were hospitalized. Forty-five hospitalized patients lived in New York City, with thirty-two in Queens, nine in the Bronx, one in Manhattan, and three in Brooklyn. One of the Queens patients was a tourist from Canada who visited Queens in early September and fell ill during the flight home. Fourteen more confirmed cases were hospitalized in the NYC suburbs, with eight in Westchester and six in Nassau County. Seven people died, including five in NYC, one in Westchester, and one in Nassau.

Over the winter, the NYC health department continued to monitor the after-effects of the outbreak and prepare for what might happen the following year. A human serosurvey in northern Queens conducted in association with the CDC indicated that about 2.6 percent (or between 1.2 and 4.1 percent) of the 46,220 people in the surveyed area had antibodies in their blood indicative of past infection with WNV. This meant that between five hundred and two thousand people in the surveyed area (corresponding to about nine thousand in the whole NYC outbreak area) had been exposed to the virus through mosquito bites, and that the sixty-two confirmed cases represented only the "tip of the iceberg," the most severe manifestations of disease. Many of the people who gave blood for the survey reported more concern about the possible ill effects of pesticide-spraying than about the possibility of infection with WNV— an attitude that mystified Hauer and Giuliani.[7] In fact, malathion killed off many insects besides mosquitoes, including butterflies, moths, and bees needed to pollinate trees and plants, and possibly also sea life. A class-action suit claiming that pesticides washed into Long Island Sound by Hurricane Floyd had caused a mass die-off of lobsters was not settled until April 2008. The timing of the die-off (September–October 1999) was suggestive, but the scientific evidence was inconclusive.

The NYC findings were roughly consistent with findings from previous studies in other countries suggesting that one in five people infected with WNV develops a mild illness with fever and headache (called "West Nile fever"), while about one in 150 infected people (usually the elderly) develop the symptoms seen by Asnis in Queens: encephalitis, muscle weakness, and mental confusion ("West Nile severe disease" or "West Nile neuroinvasive disease"). However, the sixty-two cases in NYC were unexpectedly severe. Fifty-nine of the sixty-two patients had West Nile neuroinvasive disease, and many of those who recovered were ill for as long as two months. Moreover, many remained weak or had difficulty concentrating after eighteen months.

What would happen next year? Once a zoonotic or arboviral pathogen enters the local insect and wildlife populations, it can be nearly impossible to stop. Although mosquitoes stop flying with the first autumn frost, they can survive freezing weather through a form of insect hibernation called "overwintering," in which they cling to protective surfaces such as the underside of ledges and pipes. PCR-testing of thousands of hibernating *Culex* mosquitoes in New York City detected only one WNV-positive mosquito, but left open the possibility that a hibernating WNV-positive mosquito might awaken in the spring and lay WNV-infected eggs. Intensive mosquito surveillance and control efforts after winter's end and during the following spring probably contributed to the smaller number of severe human cases (eighteen) in New York City in 2000, most of which were in the borough of Staten Island.

However, WNV quickly spread far beyond the New York area, apparently carried by birds (in addition to mosquitoes), who spread the virus along the Atlantic flyway and inland westward year by year, reaching all the West Coast states by 2003. WNV also spread south to Latin America and north to Canada, becoming endemic throughout the Western Hemisphere. Today, it is a leading cause of epidemic encephalitis in North America, and mosquito control has been re-instituted or enhanced in many localities.

In January 2000, the CDC Fort Collins laboratory issued public health guidelines for surveillance, prevention, and control of WNV infections and established a National West Nile Virus Surveillance System—now a component of ArboNET[8]—that monitors WNV in humans, birds, mosquitoes, horses, and sentinel chicken flocks. (Chickens are easily infected but do not fall ill, while horses can be severely affected. An outbreak that affected about twenty-five horses on Long Island in August and September 1999 was also traced back to infection with West Nile virus.) Screening for West Nile was instituted at blood banks and transplant centers after WNV was detected in donated blood and in transplanted organs.[9] Human cases of WNV infection in the United States rose to a peak in 2003, when 9,862 severe cases were reported; the numbers fell to 2,500 to 4,200 per year until 2012, when the United States experienced a major outbreak affecting thirty-eight states, with three-fourths of the cases reported in Texas, Mississippi, Louisiana, South Dakota, and Oklahoma.[10]

So where did WNV come from and how did it end up in New York City? Perhaps the most likely explanation is that it arrived by plane at La Guardia Airport (about four miles from College Point) or JFK Airport (fewer than ten miles away). WNV could have traveled inside a mosquito that "hitch-hiked" on a flight from a place where the virus circulates in the local bird population, such as southern France, India, or Israel. (In 2001, Asian tiger mosquitoes—the vector of dengue hemorrhagic fever—arrived in California in the cargo hold of an airplane, inside shipping boxes containing a Chinese plant called "lucky bamboo.") Or, the virus could have traveled inside an exotic bird that was smuggled into the city, avoiding both quarantine and inspection. Another possibility is that WNV arrived inside a human traveler.

The theory that the virus came by plane—inside a bird, mosquito, or person—from an area in which it is endemic is bolstered by the close match between the 1999 NYC strain of WNV and the strain isolated from the Israeli goose in 1998 (when there were no human cases in Israel) and between the NYC virus and human WNV isolates obtained during an Israeli outbreak in 2000. However, it is also possible that WNV was carried into Queens by a migrating bird. A map of migratory routes of common New World birds suggests that New York serves as an avian as well as a human travel hub, crossed by bird routes that go north into Canada and south into Mexico and many parts of South America. Although most birds migrate north and south (rather than east and west across the Atlantic), it is possible that WNV could have entered a different part of the Americas by plane, boat, or bird at some unknown time in the past, traveled to Queens by bird, and only revealed itself when favorable circumstances—a weather-induced increase in mosquitoes in a densely populated urban area—led to a small outbreak. In fact, the NYC outbreak was so small that it might have gone undetected if an astute physician in Queens had not put two and two together and called her local health department. So it is certainly possible that other small undetected human WNV outbreaks had previously occurred in other places. Arguing against this theory is the absence of reports of unexplained bird die-offs during the 1990s that would likely have occurred around the same time as a human outbreak (the basis of the "dead bird surveillance system").[11]

One more theory—that the WNV was released into New York by terrorists—has been raised and discredited several times. Early on, Layton and Fine explained to the local FBI agent in NYC that the presence of mosquito vectors and the distribution of cases were consistent with a natural SLE outbreak. Moreover, SLE, with its generally mild symptoms and low fatality rate, seems a poor choice for a bioweapon. However, the idea was raised again by Richard Preston, the author of *The Hot Zone* (a best seller about Ebola hemorrhagic fever), after the diagnosis changed from SLE to West Nile fever. Writing in the October 18 issue of the *New Yorker*, Preston drew attention to a book excerpted in a London tabloid in April 1999 in which the author—who claimed to be a former bodyguard and body-double for Saddam Hussein—asserted that a highly virulent strain of WNV was part of Saddam's bioterrorism arsenal. Despite the strange timing, with the book appearing about six months before the NYC outbreak, neither intelligence officials nor arms inspectors turned up any public evidence that Iraq had tried to weaponize WNV.

The possibility of bioterrorism was re-visited three years later, in fall 2002, during the run-up to the U.S. invasion of Iraq. (By this time, Layton had developed a close working relationship with a local FBI agent, due to 9/11 and the anthrax incidents.) An article in *Business Week* reported that in 1985, when the U.S. and Iraq were allied against Iran, the CDC had shipped samples of West Nile virus to medical researchers at the University of Basra. However, when biologists at the CDC compared the sequence of the strain sent to Iraq in 1985 to the 1999 NYC strain, they were relieved to find that the two strains were not alike.

Lessons Learned

A major lesson of the 1999 WNV outbreak is that sometimes epidemiologists need to ignore the old medical adage: When you hear hoofbeats, think horses, not zebras. SLE was an unusual diagnosis for New York City, and West Nile encephalitis was stranger still. Although exotic pathogens have traveled throughout history, attacking non-immune populations and spreading havoc in their wake, microbial travel is easier and faster today because of air travel, urbanization, and increased human contact with animals and insects that live in rain forests and other wilderness areas. Also, we now have the molecular tools to confirm that something is new or rare.

Fine says that another lesson—or maybe the same lesson put another way—is that sometimes it's good for an epidemiologist to be somewhat naïve, so that he or she is not blinded by traditional ways of thinking about a particular disease (e.g., that avian encephalitis viruses don't kill birds, that West Nile virus only causes mild illness, or that it is not worth including exotic viruses in a test panel). Perhaps it's a good thing that most epidemiologists are generalists who investigate many different types of infectious outbreaks and may not be as aware of (or invested in) current dogma as a specialist might be.

Layton and Fine agree that the biggest take-home lesson of the 1999 WNV outbreak is the need for public health and animal health officials to work together and share information on a regular basis. Due to the lack of communication, Layton and Fine conducted a complex outbreak investigation, potentially affecting millions of New Yorkers, while remaining unaware of important clues developed by veterinary colleagues. In retrospect, the delay did not have major consequences in New York City, because the diagnosis of SLE set in motion mosquito control measures that were also effective against WNV and because clinical care for SLE and West Nile encephalitis is the same. Moreover, it is likely that the SLE testing discrepancies would have eventually led to a reconsideration of the diagnosis (e.g., via Lipkin's work) even in the absence of the veterinary investigation. As Stephen Ostroff, a high-level CDC official, explained to a *New York Times* reporter, confusion is a completely normal part of a public health investigation, which can take many twists and turns before everything becomes clear.[12]

Nevertheless, the delay in making a connection between the bird deaths and cases of human disease did delay recognition that human cases were occurring beyond New York City. Areas outside of the city that experienced significant bird die-offs—including Westchester

County, Nassau and Suffolk counties in Long Island, and parts of Connecticut and New Jersey—did not institute mosquito control efforts until the end of September or early October. Two of those areas had confirmed human cases (nine in Westchester and six in Nassau).

As many public health experts have noted,[13] the great majority of emerging diseases discovered in recent years have been caused by zoonotic pathogens (like severe acute respiratory syndrome [SARS]) or by human pathogens that arose from animal reservoirs (like HIV/AIDS). It is therefore imperative that public health authorities keep abreast of outbreaks and health issues not only in humans but also in livestock, wildlife, zoo animals, and pets.[14] Fortunately, this lesson has been taken to heart by public health and veterinary professionals who advocate an integrated approach to the protection of human and animal health.[4] Today, the NYC health department has a public health veterinarian who is charged with outreach to the animal health community, and federal, state, and local public health laboratories are officially linked with high-level agricultural, wildlife, and military reference laboratories through the Laboratory Response Network (LRN), which was founded by the CDC in partnership with the Association of Public Health Laboratories and the FBI. Conceived in 1998 as part of the national bioterrorism preparedness effort, the LRN has proved to be just as important, if not more so, in responding to naturally occurring outbreaks as to human-made ones, just as OEM proved invaluable to the West Nile response, although the impetus for developing OEM was bioterrorism.

This is not surprising, because all outbreak investigations require the same epidemiologic and laboratory skills—and control efforts require the same logistic and financial resources—whether the cause of the outbreak is a well-known virus, a new or drug-resistant bacterium, or a weaponized pathogen. But bioterrorism looms large in the public imagination, and fear of bioterrorism is a greater spur to public health investment than the fear of a naturally occurring illness, such as an exotic arboviral disease that has found an ecologic niche in the Western Hemisphere. From a public policy point of view, therefore, the West Nile outbreak illustrates that investments in bioterrorism preparedness can be especially valuable if states and cities are allowed to use them flexibly to address whatever public health crisis may arise.

2

The McConnon Strain

Patrick J. McConnon, the U.S. CDC Regional Southeast Asia Coordinator for Refugees, was sick to his stomach, dreading the decision he had to make. It was a rainy day in spring 1982, and he was sitting in the CDC regional office in Bangkok, Thailand. He had expected a routine phone call, letting him know how many Cambodians—men, women, and children—were scheduled to undergo medical screening as part of the standard process of resettlement in the United States. But this time, the transport arrangements had changed. Due to the large number of refugees—nearly twenty thousand people—the medical processing would not take place nearby, at the local transit center in Phanat Nikhom. Instead, the refugees would be bused from the Kamput refugee camp directly to the Bangkok airport and then flown out of the country to a much larger processing center in Bataan, in the Philippines.

McConnon was afraid that this change in logistics might have huge unintended consequences that would be difficult or impossible to control. Just the other day, he had learned that some Kamput refugees might harbor an unusual infectious disease: a drug-resistant, life-threatening form of malaria. Knowing how easily disease can spread, he did not want the refugees to leave the country without being screened. But what should he do? If he stopped the flight to Bataan, he would prolong the misery of thousands of desperate families eager to resettle and start a new life—and there would also be major political repercussions, not to mention personal repercussions for the bearer of bad news. But if he let the refugees leave, he could be aiding and abetting the spread of a dangerous disease that could bring illness and death to thousands of people throughout the world.

McConnon pushed himself out of his chair and looked out the window at the empty daytime streets of Patpong, the gaudy red-light district of Bangkok,[1] now dark and dreary from constant rain. He thought about the homeless Cambodians in Kamput, uprooted by the Khmer Rouge, forbidden by Thai authorities to live outside the camp, and afraid to return to Vietnamese-occupied Cambodia. Then he thought about the people he worked with in other Southeast Asia countries—Filipinos, Vietnamese, Laotians, Indonesians, Malaysians— and how quickly diseases can spread from one country to another. He vowed not to let it happen on his watch.

Refugee Camps and Medical Screening

People stranded in refugee camps, displaced, impoverished, and malnourished, are at special risk for infectious diseases such as malaria, measles, and cholera that flourish in crowded and unsanitary living conditions. When infected refugees are moved to new holding sites, repatriated, or resettled in new countries, they can bring these diseases with them. As a result, public health officials like McConnon have overlapping and sometimes conflicting aims: to safeguard the health and welfare not only of the refugees themselves, but also of the people in countries that host refugees camps or accept refugees as permanent residents

The spread of smallpox after the 1971 Pakistani civil war illustrates what can happen when a pathogen incubated in a refugee camp infects the wider population. Smallpox was carried to the newly established nation of Bangladesh by refugees returning home from India. According to public health lore, the presence of smallpox in the camps was detected by an epidemiologist in Atlanta, sitting in his living room watching TV, who noticed a man with a suspicious rash in a newsreel about a camp near Calcutta. The man's face was covered with the large pus-filled lesions characteristic of advanced smallpox. The epidemiologist called the director of the CDC, who called the director of the World Health Organization (WHO) Smallpox Vaccination Program, who called the Indian Ministry of Health. But it was already too late. Thousands of Bengalis had already left the camp, leading to widespread outbreaks in Bangladesh and making it the last Asian country to finally eliminate smallpox.

Medical screening procedures to prevent refugees from carrying disease into the United States were written into the Refugee Act of 1980, which was drafted in response to the refugee crisis that followed the end of the Vietnam War and the rise and fall of the Khmer Rouge in Cambodia. The Refugee Act requires applicants for resettlement in the United States to pass a rigorous medical examination that includes testing for "inadmissible" or "quarantinable" diseases, such as smallpox, cholera, plague, diphtheria, infectious tuberculosis, yellow fever, and viral hemorrhagic fevers like Ebola and Marburg. Severe acute respiratory syndrome (SARS) was added in 2003, and influenza caused by viruses with the potential to cause a pandemic was added in 2005.

For the Cambodian refugees who applied for U.S. visas in 1982, medical screening was the last step in a long process that began with a preliminary interview by a worker from the Joint Voluntary Agency (JVA), a nongovernment organization under contract to the U.S. Department of State. Typical JVA workers were a mix of idealistic young Americans, hoping to help the refugees, and less altruistic individuals that the old-timers called "world travelers" or "WTs"—adventurers and drifters hoping to earn money to fund their peregrinations. The job of the JVA interviewer was to ensure that all prospective U.S. immigrants fulfilled the minimum requirements of the U.S. Immigration and Naturalization Service (INS). They gathered as much information on each applicant as they could to gain insights into their local customs and attitudes. They asked each applicant whether he or she had relatives or sponsors in the United States or had ever been employed by the U.S. government. They also tried to determine whether the applicant had been a member of the Khmer Rouge. Members of Cambodian resistance groups, including those faithful to Pol Pot, to Prince Norodom Sihanouk, or to former Prime Minister Son Sann, continued to congregate on the Cambodian side of the border and sometimes used the border camps as bases for their operations.

Individuals recommended for INS consideration by JVA workers were interviewed a second time by INS personnel and then—if they passed—were sent to Phanat Nikhom, about sixty-five miles southeast of Bangkok, for medical screening. Many of the refugees were in poor health, requiring treatment for malaria, tuberculosis, anemia, hepatitis, upper respiratory tract infections, or intestinal parasites. Individuals with malaria were treated with a dual regimen of quinine and tetracycline, rather than the standard local treatments of chloroquine (CQ) or sulfadoxine-pyrimethamine (SP), because many local malaria strains were either CQ-resistant or SP-resistant. Individuals with tuberculosis were not allowed to leave for the United States until their illness was controlled with a multidrug treatment regimen administered for at least six months.

Malaria—a scourge known since ancient times—is even today one of the primary causes of illness and death in refugee camps. However, it is not included on the U.S. list of inadmissible diseases or on the WHO list of reportable diseases because it is not transmissible from person to person and is not rare or unusual. It is endemic in the equatorial regions of Asia, Africa, and South America, where it currently affects 350 to 500 million people. Malaria was eliminated in most urban areas in Cambodia and Thailand after World War II due to the WHO Global Malaria Eradication Campaign (1955–1969), whose main weapon was the insecticide DDT because there was (and is) no vaccine against malaria. By the time of the Khmer Rouge, however, malaria had resurged in rural areas. In 1979, it was the most common cause of death among the Cambodian refugees who fled through western Cambodia to reach the border with Thailand.

The Investigators

McConnon, born and raised in Minnesota, was an expert in public health operations and logistics who had worked as a public health advisor in Ohio, Virginia, and Minnesota, and at CDC headquarters in Atlanta, as well as with the WHO Smallpox Eradication Program in Bangladesh and Somalia. He had arrived in Bangkok nine months earlier, in August,

1981, with his wife, Kate, young daughter, and teenage son. His role included quality control and oversight of refugee medical screening, as well as evaluation of any acute public health issues that affected people displaced in the aftermath of the Vietnam War. These included refugees from Cambodia, Vietnam, and Laos (including Hmong) who fled over-land to Thailand, and the Vietnamese "boat people" who fled by sea to any country that would admit them, ending up in refugee camps in Thailand, the Philippines, Singapore, Hong Kong, Malaysia, or Indonesia.

McConnon visited the refugee camps in these countries every two months, staying in the tin-roofed huts or guest houses that served as staff quarters in each camp. He conferred with the relief workers who staffed the camps' medical clinics and facilitated movement of refugees to processing centers for medical screening and classes in language and cultural orientation. In 1982, the United Nations High Commissioner for Refugees (UNHCR) supported three major international processing centers, in Phanat Nikhom, Thailand; Galang, Indonesia; and Bataan in the Philippines. Refugees seeking resettlement in Western countries were also processed and screened at transit camps in Singapore, Hong Kong, Macao, and Malaysia.

While in Bangkok, McConnon lived with his family in a tradi-tional wooden Thai house within a large compound that also included a three-story apartment building with a Japanese family on each floor, as well as a central building—the "main house"—that provided lodging for the compound's staff, including housekeepers, cooks, and gardeners. The McConnons' house was all red inside, with mahogany floors and ceilings, more cozy than elegant, but old and impressive, with air conditioning in the upstairs bedrooms and ceiling fans in the living and dining rooms. Air conditioning was an important amenity because temperatures in Bangkok rarely dropped below 70 °F (21 °C), even during the "cool season" (December through February). During the rainy season (May through November), the whole compound would be flooded for days. After breakfast, McConnon and his wife and children would take off their shoes and socks, roll their pants or skirts up to the knees, and wade through the water to the road outside the compound to catch a taxi or board the school bus.

McConnon's children attended an international school in Bangkok, and his wife Kate worked at the Orderly Departure Program (ODP). ODP was established in 1979 as a safe and legal path to emigration for Vietnamese nationals with family in the United States, including children born to Vietnamese mothers and American G.I. fathers. While all of the McConnons learned to speak Thai well enough to say hello to neighbors and order food in restaurants, McConnon's ten-year-old daughter became quite fluent. On school holidays, the McConnons' cook and housekeeper took her with them to the countryside to attend Buddhist religious festivals, which were colorful and exciting and unlike anything she had experienced at home.

Among the McConnon's closest friends in Bankgok were two relief-agency colleagues, Roland Sutter, a Swiss physician, and his Vietnamese wife, Xuan, a former refugee who fled Saigon by boat with her family in 1975. Rescued by the U.S. Seventh Fleet, she had lived in a camp in Guam, resettled in Canada, and returned to Southeast Asia as a Canadian relief worker at the Galang refugee camp in Indonesia. Her current job was as an interviewer and translator at ODP, where she worked with Kate McConnon. Her husband, whom she had met in Indonesia, was the regional medical officer of the Intergovernmental Committee on Migration, an organization founded in Switzerland in 1951 to facilitate emigration of Europeans displaced during World War II. In 1982, the Intergovernmental Committee on Migration was on contract to the U.S. Department of State to oversee the doctors and nurses hired to screen U.S.-bound refugees from Southeast Asia. A fit, sophisticated man in his thirties, Sutter was responsible for organizing and overseeing medical screening for all refugees in Southeast Asia. His duties included verifying that the medical screening of refugees accepted for resettlement in a given country fulfilled the legal screening requirements of that country. In Phanat Nikhom, he supervised the medical personnel on contract to the U.S. Department of State and worked in partnership with McConnon to ensure fulfillment of U.S. government regulations.

McConnon's other duties required a certain degree of flexibility and resourcefulness, especially when he was called on to do odd jobs related to health but outside his usual job description. Once, when he was still new at the job, his secretary informed him that he had a "dog head problem." In the waiting room was a man from Phanat Nikhom clutching an oversized red plastic ice bucket from which protruded the elongated snout of a very large dog. Fascinated and repulsed (and lacking any expertise in rabies diagnostics), McConnon carried the bucket by taxi to a rabies specialist at Chulalongkorn University who confirmed that the head had belonged to a rabid dog and that any persons bitten by that dog required treatment. Another time, McConnon helped a senior Thai dignitary suffering from respiratory distress by shipping a vial of his blood to CDC laboratories in Atlanta, which detected *Histoplasma capsulatum*, a fungus spread by bird or bat droppings that can cause a fatal lung disease if left untreated.

McConnon's job also required a willingness to take a stand. Although the Bangkok posting was his first assignment in refugee health, McConnon had seen in Bangladesh first-hand evidence of international disease spread by infected refugees. Tall and bearded, McConnon is a soft-spoken, straightforward person who is not afraid to express his mind. Having taken on many challenging assignments for the CDC, he was well aware of the political pressures and personal agendas that can interfere with public health decision-making, and he tried always to focus on the public good. Since arriving in Bangkok, he had twice ordered the cancellation of flights carrying Vietnamese refugees to the United States. The first occurred in Hong Kong, where a chickenpox outbreak was raging among young children in the refugee camp. The second was in Singapore, when procedures for preboarding screening for fever and rash were poorly implemented by inexperienced airport personnel. McConnon had agonized over these difficult but necessary decisions.

After the airport incidents in Hong Kong and Singapore, McConnon and Sutter worked together to establish routine procedures for preboarding screening of refugees for fever, cough, and rash. Another joint project involved working with refugee camps and transit centers in Southeast Asia to institute directly observed therapy for tuberculosis, the same strategy of daily, documented treatment used in the United States to contain the reemergence of tuberculosis in New York City.

To keep abreast of any new health issues that might arise in the refugee camps, McConnon and Sutter attended monthly UNHCR coordination meetings attended by representatives of more than 50 humanitarian relief agencies that worked in the refugee camps along the Thailand–Cambodia border. The meetings were run by Arcot G. Rangaraj, the UNHCR health officer in Thailand, a former Indian Medical Service surgeon with an impressive military record and a ramrod-straight bearing. In 1945, as the Regimental Medical Officer of the 152nd Indian Parachute Battalion, he had parachuted into Burma (now Myanmar) at Elephant Point as part of an attack on Japanese-occupied Rangoon called Operation Dracula. During the Korean War, he commanded the Indian Army's 60th Parachute Field Ambulance Platoon, a mobile army surgical hospital that evacuated wounded British troops and was known for its courage under fire. Skilled at both medicine and war, Rangaraj joined the WHO in 1969 as Senior Advisor on Smallpox Eradication, first in Afghanistan and later in Bangladesh and the Arabian Peninsula. McConnon and Rangaraj developed a special bond when they realized they had both worked in Bangladesh in 1975, the final year of the smallpox eradication effort in Asia.

Rangaraj's UNHCR meetings served as a relief-agency grapevine for refugee health issues. There health workers learned about vaccine-preventable neonatal tetanus among the Hmong—who refused to allow their sons to be vaccinated because they believed it caused a baby's testicles to recede—as well as malnutrition, outbreaks of enteric disease, and bites by rabid animals. (As McConnon found out after his "dog head" experience, canine rabies was fairly common in Thailand, and not only in the refugee camps, because the local custom was to feed wild dogs rather than kill them or house them in dog pounds).

It was at a UNHCR meeting that McConnon and Sutter first heard about cases of malaria in Kamput that exhibited a type of multidrug resistance unknown outside of Cambodia and Thailand (and extremely rare in Thailand). According to healthcare workers at the Kamput clinic, the patients did not recover after treatment with either

CQ (the cheapest and most widely used antimalarial drug) or with SP. McConnon made a mental note to keep track of malaria patients in Kamput in case additional cases of multidrug resistance turned up during the following months.

The Public Health Problem

The refugee situation in Southeast Asia involved hundreds of thousands of people uprooted by a succession of wars and invasions. McConnon arrived in Bangkok at the end of the second large wave of refugees who fled into Thailand. The first wave (1975–1977) occurred after the fall of Saigon and the withdrawal of U.S. troops from Vietnam. The second and largest wave (1978–1982) occurred after the defeat of the Khmer Rouge by the Vietnamese at a time of widespread famine. A third wave occurred in the mid-1980s after Vietnam expelled its ethnic Chinese population in reaction to an invasion by China that led to continued regional conflict.

The second wave included about two hundred thousand Cambodians who reached Thailand after surviving malnutrition, exposure to malaria, and gunfire from soldiers, bandits, and border guards. They included members of Pol Pot's army, as well as starving civilian families fleeing the occupying Vietnamese forces or the Khmer Rouge, or both. UNHCR supported construction of makeshift border camps, and the World Food Program provided food and humanitarian aid, assisted by the United Nations Children's Fund (UNICEF), the International Committee of the Red Cross, and many other relief agencies.

By 1982, the Thai border camps had evolved into small villages with bamboo-and-thatch–roofed huts, vegetable gardens, food markets, and medical and relief services. Certain camps, including Kamput, were designated as holding centers for refugees awaiting repatriation in Cambodia or resettlement in a new country. The largest camp, Khao I Dang, housed about 130,000 refugees, making it the largest Cambodian city in the world at the time, because Phnom Penh had been "evacuated" by the Khmer Rouge.

About twenty thousand second-wave Cambodians who claimed refugee status were admitted to the United States from 1979 to 1981; most had family ties or economic sponsors in the United States. In mid-1981, however, the INS ruled that the Refugee Act of 1980 required case-by-case proof that a refugee was at risk for persecution if repatriated in Cambodia. This requirement was difficult to fulfill, and many previously eligible visa applicants were rejected by INS and left in stateless limbo in the border camps. By this time, however, the refugee crisis in Southeast Asia had become a political issue in the United States because of widespread news reports about the Vietnamese boat people and the "killing fields" of Cambodia. (Closer to home, the 1980 Mariel boat lifts—which included individuals released from Cuban prisons—also heightened public awareness of refugee issues.) Several U.S. aid groups and individuals, including State Department officials, senators, and congressmen, spoke out on behalf of the Cambodian refugees and acknowledged U.S. responsibility in creating the refugee crisis. They drew attention to failed attempts at forced repatriation of Cambodians by the Thai government that resulted in the death of refugee families caught between the occupying Vietnamese army and Cambodian resistance forces.

In early 1982, the INS agreed to relax the standards of proof and admit about twenty thousand additional refugees, all at one time, as a step toward emptying the border camps. This decision was a welcome development for everyone. It meant that thousands of uprooted families might soon find homes in the United States and that the border camps might soon be closed.

However, there was one problem, obvious to a public health person like McConnon, if to no one else. To speed the processing of so many people, the State Department decided to move thousands of refugees from Khao I Dong to Kamput for prescreening and then (if JVA approved them) directly to the refugee center in Bataan, bypassing the usual six to eight week stay in Phanat Nikhom. Medical screening would not take place until after the refugees arrived in the Philippines.

From a public health point of view, moving unscreened refugees out of a country is never a good idea, and McConnon thought it was especially unwise in this instance because of the multidrug-resistant malaria reported in Kamput. If refugees from Khao I Dong acquired drug-resistant malaria during their stay in Kamput—which was entirely possible—they could spread disease to the Philippines and beyond. Although refugees with obvious malaria symptoms, like fever and chills, would be prevented from flying by the preboarding screening procedures instituted by McConnon and Sutter, people who were infected but not yet experiencing symptoms could go undetected. (The incubation period—the time between infection and the appearance of symptoms—for malaria is usually ten days to four weeks.)

To make matters worse, the malaria cases reported at the Kamput clinic were caused by *Plasmodium falciparum* malaria, the more dangerous of the two types of human malaria endemic in Southeast Asia (the other is *Plasmodium vivax*).[2] Malaria is caused by parasites that are carried by mosquitoes and cause severe fever and chills during the "blood-stage" portion of the parasite life cycle (see insert). *P. falciparum* is especially dangerous to children under five years old, who are too young to have developed immunity through repeated exposures, and to pregnant women, who are at high risk for anemia. Coma and death can result from cerebral malaria, a complication that occurs when *P. falciparum*–infected blood cells attach to the walls of blood vessels in the brain.

To make matters even worse, the Bataan Peninsula was a particularly malarious area. Forty years earlier, malaria had been a significant factor in the Allied defeat at the Battle of Bataan. Thousands of Filipino and U.S. troops died from malaria before and during the battle, as well as during the forced "Death March" to a prison camp in Capas, Philippines, that followed the surrender to the Japanese. A U.S. veteran recalled Bataan as "one of the most heavily malaria-infested areas in the world,"[3] with mosquitoes everywhere and a scarcity of the antimalarial drug quinine for infected soldiers on both sides of the war. During the 1980s, despite the partial success of the WHO Global Malaria Eradiation Campaign, malaria was still an important health problem in several Filipino provinces, including

Bataan. The introduction of a multidrug-resistant strain into Bataan would be disastrous.

Although McConnon did not know it, the Thailand–Cambodia border region was a longtime incubator of drug resistance in *P. falciparum* parasites. In fact, Kamput is located within twenty-five miles of Pailin, a Cambodian mining town identified by the medical historian Randall Packard as a world epicenter for the development of antimalarial drug resistance.[4] Packard argues that resistance to CQ—the drug that replaced quinine as the antimalarial drug of choice after World War II—arose in Pailin during the 1950s and 1960s from a confluence of factors that included mining practices, migration patterns, and misguided public health efforts.[5]

In 1982, Pailin was known as a Khmer Rouge stronghold, a place where Khmer Rouge leaders retreated after the Vietnamese Army overthrew Pol Pot, funding resistance operations by smuggling gems and timber. But in the aftermath of World War II, Pailin's mines had attracted a constant flow of transients—mostly impoverished farmers—from other parts of Cambodia and from Thailand, Vietnam, and Myanmar, who dug up rubies and sapphires and sold them for a few dollars apiece to local merchants. The migrants slept out in the open or in rudimentary shelters with no bednets to protect them from *Anopheles dirus*, a local night-biting mosquito that bred in used-up mining shafts full of stagnant water. Although people who survive to adulthood in malarious areas usually have partial immunity to malaria, many of the migrants came from less malarious areas, had little or no immunity, and tended to become very sick, with high concentrations of parasites in their blood that were transmitted by mosquito bite to other people.

Since *A. dirus* is an outdoor feeder unaffected by indoor spraying with DDT, efforts to control malaria among the miners focused on administration of CQ as a preventive measure. But the procedure apparently backfired. Although the means of distribution varied from year to year—given in pill form in varying doses or ingested in food seasoned with medicated salt—the dosages were never high enough to kill any but the most CQ-sensitive parasites, providing a selective advantage to parasites that were CQ-resistant. Moreover, migrants suffering from malaria who used their mining profits to purchase additional CQ usually could afford only enough for a subcurative dose, further boosting the survival rates of CQ-resistant parasites. Successive malaria epidemics among new groups of nonimmune migrants who ingested high but noncurative doses of CQ apparently amplified CQ resistance from year to year until, by the early 1970s, most *P. falciparum* parasites in the border region were moderately or extremely resistant to CQ. SP, a new drug combination, came into use in Thailand in the mid-1970s as a second-line drug for use when CQ failed.

In 1982, malaria resistance to both CQ and SP (dual resistance) was rare and had never been reported outside of Thailand and Cambodia—and McConnon wanted to keep it that way. He thought about his decisions to cancel the flights in Hong Kong and Singapore, both of which had been very difficult. He was now faced with the prospect of exercising his flight-cancellation authority a third time, in a situation that involved many more people and a much more highly charged political situation.

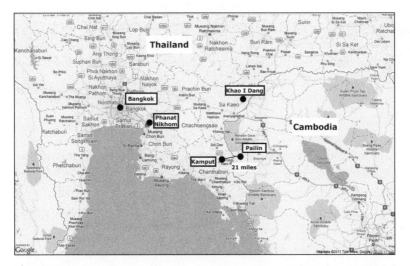

Map of Thailand and Cambodia indicating the locations of the Kamput and Khao I Dang refugee camps, the transit center at Phanat Nikhom, and the mining town of Pailin. *Source:* CSTE: Edward Chow, Lauren Rosenberg, and Jennifer Lemmings.

McConnon confided in Sutter, who agreed that multidrug-resistant malaria must not spread on their watch. McConnon recalled his experiences in Bangladesh battling smallpox spread by repatriated refugees. He and Sutter joked bitterly about the world-renowned, drug-resistant "McConnon strain" that might spread worldwide if they failed to act.

With Sutter's encouragement, McConnon contacted Mac Allan Thompson, the State Department official in charge of the U.S. Indochina Refugee Program. Thompson, a graduate of the Colorado School of Mining and Engineering, was a Vietnam veteran who had returned to Southeast Asia as a government official. He was energetic, athletic, and fearless and had only recently returned to work after breaking his back while teaching sky-diving—his weekend hobby— to Thai military officers. He was also thoughtful and independent- minded, with the air of someone who was always one step ahead. He had contacts and sources of information everywhere, including such remote places as the Golden Triangle, an opium-producing area in the mountains of Myanmar, Laos, and Thailand.[6]

McConnon was somewhat nervous. Thompson could be outspoken and play rough if someone contradicted him or interfered with his plans and operations. He also knew that Thompson was under intense pressure to move the refugees to Bataan. With great trepidation—and the specter of the "McConnon strain" filling his mind—McConnon took Thompson aside and said he had something surprising and serious to tell him. Then he explained why they could not send unscreened refugees to the Philippines.

Thompson's first reaction was to dismiss McConnon's concerns out of hand. He reminded McConnon that resettlement of the refugees was important from both humanitarian and political standpoints and that additional supplies had already been shipped to Bataan. When McConnon insisted he would not allow the refugees to board the air- plane, Thompson reminded him that he had clear authority only to prevent travel into the United States, not necessarily between Thailand and the Philippines.

When McConnon still did not back down, Thompson asked whether McConnon was absolutely certain of the accuracy of the reports of malaria from the Kamput clinic. After all, the clinic lacked sophisticated diagnostic tests, and malaria can be mistaken for other febrile diseases. McConnon admitted he was not certain. Nevertheless, he knew from experience that when it comes to public health, being over-cautious is a good thing. "Well then," said Thompson, "we have four weeks before the move to Bataan. If you can provide some evidence that the danger is real, I'll see what I can do."

The Investigation

McConnon and Sutter lost no time in mounting an investigation to confirm the reports of multidrug-resistant malaria and determine how the disease was spread. Although neither McConnon or Sutter had formal training in epidemiology, they were familiar with basic epidemiologic tools, such as case-control studies and retrospective studies, that could help them figure out why some people were getting ill and others were not. If they could identify behaviors or activities that put refugees at risk for acquiring malaria (called "risk factors"), they would know which individuals were most likely to be infected with the drug-resistant disease. In that instance, they could stop those individuals from going to Bataan, and McConnon might not have to cancel the flights.

They began by brainstorming about occupations or activities that might bring the refugees into contact with mosquitoes. Although they did not know which *Anopheles* species lived in eastern Thailand, they assumed that mosquitoes are likely to be highly concentrated in forests or near swamps or other bodies of standing water where they breed. They also knew that some types of mosquitoes are most active at night or evening, whereas others tend to feed in the early morning or late afternoon.

The Kamput refugee camp was tiny—perhaps ten acres—and entirely surrounded by barbed wire. It included a living area filled with family huts, a forested area, and some swamplands. Some refugee families kept chickens in their yards and had planted small vegetable gardens near their huts or near the forest. It was possible that farmers whose gardens bordered forests were most at risk for malaria.

Or perhaps other tasks—such as fetching water from wells or ponds in the early morning or taking garbage to be buried in a makeshift dump inside the forest—increased proximity to mosquitoes.

To test these hypotheses, McConnon and Sutter designed a case-control study. This type of study compares a group of patients (the case-patients) with a group of people who are not sick (the controls) to determine which risk factors are associated with illness and which are not. They planned to compare malaria patients seen at the Kamput clinic with healthy camp residents matched by age and sex. To confirm the case-patients had malaria and not some other febrile disease, they needed to perform standard diagnostic tests for malaria called "thick smears" and "thin smears," which they did not know how to do. They called on Dr. Peter Echevarria, a malaria expert at the U.S. Armed Forced Research Institute for Medical Science in Bangkok, and convinced him to lend them a hand. Dr. Echevarria not only taught them how to make blood smears but also organized the laboratory part of the case-control study.

Viewed under a microscope, a "thick smear"—a finger-prick's drop of blood spread out on a glass slide—reveals the percentage of red blood cells that are infected with malaria parasites (the parasite density or "parasitemia"). Parasitemia high enough to be detected on a thick smear is generally seen during the blood stage of the illness, when the patient is experiencing fever and chills. A "thin smear"—a drop of blood spread over a larger area—provides a clear view of individual parasites, enabling the investigator to determine which type of parasite is present (for example, *P. vivax* or *P. falciparum*).

With assistance from Dr. Echevarria and colleagues in the Kamput clinic, McConnon and Sutter obtained blood samples from Kamput residents who had malaria symptoms (fever and chills), as well as from a control group of healthy refugees. McConnon took care of the logistics, and Sutter prepared the blood smears. They determined that nearly all of the case-patients did in fact have malaria—and that each malaria case was caused by *P. falciparum* parasites. To their disappointment, however, they did not see any clear differences in activities between the people with malaria and the people without it, probably because the number of newly identified malaria cases was too small to generate any statistical correlations.

McConnon and Sutter decided to use a second epidemiologic tool, a retrospective study of malaria cases observed in Kamput over the previous six months, using the medical records of the Kamput clinic. They began by agreeing on a case definition for malaria: high fever accompanied by headache, muscle pain, chills, sweats, nausea, or diarrhea. With help from Dr. Eschevarria, they searched through the logs of the Kamput clinic, line by line, to find people in whom malaria had been diagnosed or whose symptoms matched the case definition, from December 1981 through May 1982. For each case of malaria, they recorded the patient's age, sex, marital status, and occupation. They also recorded the drug treatment each patient received and whether the treatment failed.

McConnon and Sutter plotted the location of each malaria case identified in either the case-control study or the retrospective study on a map of Kamput, hoping to see a pattern. But the data did not support any of their hypotheses. There was no correlation between risk for malaria and activities such as farming, house-building, chicken-raising, water-collection, and working near the forest, swamp, or garbage dump. In fact, the distribution of malaria cases seemed entirely random, except for one thing: nearly all cases involved males thirteen to fifty-five years old. The few malaria cases among women or children were found in the clinic records as part of the retrospective study and had not been confirmed with blood smears, so they might not have been malaria cases at all.

What did this mean? McConnon and Sutter were stumped.

The Missing Puzzle Piece

Seven days before the refugees were about to embark for Bataan, McConnon and Sutter found a solution to their dilemma in an abrupt and unexpected way. They were sitting at the bar at the local JVA guest house on the main road adjacent to the Kamput camp, debating what to do next. They had no new evidence to convince Thompson to hold the refugees in Thailand. The flights to Bataan were imminent, and the emergence of a "McConnon strain" no longer seemed an impossible joke. How could they stop this disaster?

Sutter had informed his superiors about the problem, and McConnon had alerted the CDC. But McConnon's supervisors believed that an on-the-ground issue was best evaluated on-site, rather than half a world away in Atlanta. So McConnon continued to agonize over what he should do. Should he make more phone calls to CDC scientists or public health experts to see if they could think of any alternatives? Should he try again to persuade Thompson? Should he threaten to resign his post in protest? These questions had kept McConnon up at night, and now he and Sutter were reviewing their data one last time. They must be missing something big—something difficult to see because of inadequate data, too few active cases, hypotheses they hadn't thought of, or questions they hadn't asked. There had to be something.

McConnon and Sutter ordered beer from the guest house kitchen. As they continued the postmortem of their inconclusive data, they were joined by colleagues from JVA who had spent the day in Kamput interviewing refugees. One was a Texan named Richard who played on McConnon's softball team in a recreational league run by the U.S. Embassy. Richard belonged to the idealistic category of JVA workers, who wanted to improve the lives of the refugees. "McConnon," he said, "You don't look so good. Let me buy you another drink."

"I feel like crap," said McConnon, and proceeded to tell him about the malaria investigation and how none of their theories held up. To his surprise, the men from JVA started smirking, and Richard was positively gleeful: "Why didn't you tell me about this before, you idiot! I could have told you the answer a long time ago! We have a saying at JVA that anyone in the camps with malaria is a gun-runner for the Khmer Rouge!"

McConnon's jaw dropped. Sutter gave a quick "oh yes!" smile. This was the missing puzzle piece—a convincing explanation of their data and a solution to their problem. The refugees living in Khao I Dang—the only camp directly overseen by the Thai Government and UNHCR—were unlikely to be members of Cambodian resistance groups. But Kamput itself, like each of the small border camps, had been infiltrated by resistance groups from its inception. So it was not difficult to believe that a small number of smugglers—men and adolescent boys—were crossing the border at night on their way to Khmer Rouge bases like Pailin, exposing themselves to night-biting mosquitoes and carrying disease as well as guns, sapphires, or timber between Thailand and Cambodia.

McConnon phoned Thompson to withdraw his objection to the State Department's travel plans for the U.S.-bound Cambodian refugees. He felt confident the JVA interviewers and INS officials would not recommend males with a history of recent malaria (whether drug-resistant or drug-susceptible)—or any other males suspected of being Khmer Rouge—for inclusion in the group of refugees sent to Bataan. Moreover, the Kamput Clinic continued to report sporadic malaria cases among males only, indicating that malaria transmission was not occurring within the camp, where it would have affected men, women, and children.

Much relieved, McConnon and Sutter realized that in their haste to meet Thompson's deadline they had neglected a key part of a public health investigation: talking to local people who understand the day-to-day realities of the affected population. Without that basic knowledge, interpretation of epidemiologic data can be difficult or impossible. While an intense focus on data and methodology is essential to any investigation—and ingrained by scientific training and habit—veteran epidemiologists tend to acquire what anthropologists call "participant observation" skills: the ability to gain qualitative understanding of local activities and attitudes through their own experience. Until they gain those skills, inexperienced epidemiologists can miss crucial information that (as in Poe's *The Purloined Letter*) is hiding in plain sight.

Aftermath

About three-fifths of the twenty thousand Cambodians interviewed at Kamput were recommended by JVA, approved by INS, and flown to Bataan for medical screening, language classes, and orientation. A total of 20,234 Cambodian refugees from the border camps were resettled in the United States during 1982. Many were reunited with "first wave" family members who had created Cambodian–American enclaves in towns such as Long Beach, California, and Lowell, Massachusetts. Others were resettled in groups in large cities such as Atlanta, Boston, Chicago, Cincinnati, Columbus, Dallas, Houston, New York City, and Phoenix.

In December 1982, the U.S. government discontinued financial support to the refugee camps, and Kamput was closed soon thereafter. Khao I Dang remained open for another decade, not closing until 1992. Most Cambodians who remained in the camps were considered economic migrants rather than refugees, and relatively few were eligible for resettlement in the United States or other Western countries. In 1993, the last Cambodian refugees were repatriated, and the rest of the border camps were closed. Of about 370,000 displaced Cambodians, about 220,000 had been admitted to Western countries, including 147,000 to the United States.

Twenty-five years after the closure of Kamput, we can say with confidence that U.S.-bound Cambodian refugees did not spread multidrug-resistant malaria to the Philippines. That this was a real danger and not an overreaction by McConnon and Sutter is borne out by DNA studies that identify Southeast Asia (especially the Thailand–Cambodia border region) as the origin of malaria strains with SP- or CQ-resistant mutations that subsequently spread to Oceania, Africa, and South America, especially during the 1980s and 1990s.[7] Today, multidrug resistance to CQ and SP is found throughout the Mekong region, as well as in East Africa.

In contrast, multidrug-resistant malaria has made little headway in the Philippines, where malaria has been eliminated in most of the country. Malaria is no longer found in Filipino cities and highlands and is a major cause of death in only five of eighty-one provinces (Palawan, Isabela, Tawi-tawi, Sulu, and Butuan City). Malaria still occurs in the province of Bataan but at a very low level. Until recently, CQ remained the antimalarial drug of choice throughout the Philippines, with SP reserved for treatment of CQ-resistant illness. In 2003, however, in response to increased reports of resistance to either drug alone, the Filipino Department of Health began recommending combined use of CQ and SP as first-line therapy, with artemether-lumefantrine as the second-line drug.

It seems inevitable that this change in Filipino medical practice eventually will lead to the emergence of strains of malaria with resistance to both CQ and SP. However, the consequences to the Philippines— and perhaps to other countries, if the resistant strains spread—will not be as grave as they would have been in 1982, when fewer drug choices were available to treat malaria.

3

Sorrow and Statistics

In March 1981, the coroner's office of the Province of Ontario alerted the Toronto police to four suspicious deaths in cardiology Ward 4A of the Hospital for Sick Children, a hospital that specializes in caring for babies with complex heart disease. In each case, the baby's condition had suddenly deteriorated, with erratic heartbeat, vomiting, and gagging—symptoms that suggested intoxication with digoxin, a common heart medicine that is dangerous when given in too high a dose.

The police zeroed in on Susan Nelles, a twenty-four-year-old nurse assigned to care for three of the infants who died, including Justin Cook, who had required constant nursing care. Two officers came to her house and accused her of giving baby Justin a deliberate overdose of digoxin. To their surprise, Ms. Nelles—a tiny blond woman with a quiet and confident air—neither cried nor proclaimed her innocence. Instead, she asked for a lawyer and produced a piece of paper with the handwritten names of two criminal attorneys. The police arrested her on the spot.

Over the next few weeks, however, homicide investigators made a series of additional discoveries that suggested an even greater tragedy. Working with the coroner's office, they identified many more suspicious infant deaths—perhaps as many as twenty-five or thirty—going back many months. Could all of these deaths be due to Nurse Nelles? The case against her fell apart during a preliminary hearing when the court learned that she had been off-duty during the death of Janice Estrella, one of the four babies named in the original murder charge, and on vacation during the death of a fifth child, Stephanie Lombardo, whose exhumed body contained high levels of digoxin—although digoxin had not been prescribed for her.

David Vanek, the preliminary hearing judge, was shocked that "someone was going about our beloved Sick Children's Hospital killing babies!"[1] Nevertheless, when the hearing ended, in May 1982, he declared that Nelles was an excellent nurse with an excellent reputation who should never have been arrested.[2] The news caused an explosion of agitation, grief, and excitement. If Susan Nelles, the notorious baby-killer, pilloried in the press for more than a year, was in fact innocent, what had caused the deaths of the babies at the Hospital for Sick Children? The police continued to believe that Nelles was guilty—and perhaps in cahoots with another nurse—while the doctors at the Hospital for Sick Children, including David Carver, Chief of Pediatrics, insisted it was all a terrible mistake. The doctors thought the cluster of deaths was a statistical anomaly—a large number of natural deaths occurring within a small period of time. They identified medical reasons for each successive failure to resuscitate a failing baby and reassured the nurses over and over that nothing was wrong. The doctors also discounted the high digoxin readings, citing well-regarded toxicologists who questioned the accuracy of the testing methods, especially when used on embalmed tissues or autopsy tissues.

In January 1982, while the preliminary hearing was in session, the hospital experienced a second major shock when five babies in the neonatal ward fell ill and one died of what appeared to be a contagious bloodstream infection. The hospital authorities evacuated the neonatal ward and moved the sick babies to the intensive care unit (ICU) for medical evaluation. With the hospital still under a cloud of suspicion, they decided to call in outside help. At the hospital's request, the Ontario Ministry of Health invited the Center for Disease Control (CDC) to send an officer from the CDC Epidemic Intelligence Service (EIS)—the U.S. training program for disease detectives—to lead a team of investigators appointed by the Ontario Department of Health. (As described in Chapter 1, EIS officers are medical-detective trainees who are sent out into the field to gain first-hand experience in solving real-life mysteries.)

When Steven Solomon, the EIS officer, arrived from Atlanta, the deserted hospital ward reminded him of an end-of-the world movie in which all of the people are gone but all of their things are left behind. He knew about the previous deaths and the arrest of a nurse and was aware that policemen, including members of the Royal Mounted Police, made frequent visits to the hospital. He was also aware that the hospital workers, though unfailingly polite and professional, were tense and concerned.

The Hospital for Sick Children—known affectionately as "Sick Kids"—was an august institution, a world leader in pediatric care, famous for saving children who would otherwise have died of heart disease. The idea that murder had been committed at the hospital had shaken Sick Kids to its core, dividing the staff and straining their shared sense of mission. The doctors (in those days nearly all men) were distressed and bewildered at having their competence and benevolence questioned by journalists and police detectives. The nurses were distraught over the infant deaths and the arrest of a well-respected colleague. Solomon felt especially bad for the nurses on the neonatal ward, who, in addition to everything else, were anxious about the newborns that had fallen ill while in their care. He admired the tact and skill of Evelyn Wallace, a colleague from the Ontario Ministry of Health, who soothed the worried staff and helped move the investigation forward. He was grateful that she let him be the face of the investigation, although she was more experienced than he was. He was also grateful that she worked with him on everything, discussing all key points, providing his first and finest experience of true professional partnership.

The circumstances of the two incidents were very different. The five sick babies on the neonatal ward had fallen ill within one afternoon, surrounded by staff members and visitors. The hospital had already disproved the bloodstream-infection hypothesis and ruled out the possibility that toxic fumes from repair work on the roof had entered the room through air ducts. What else might it be? Within a matter of weeks Solomon's team found a possible answer while conducting a detailed review of the deserted medical ward. They suspected that several nurses had inadvertently injected the infants with epinephrine (a hormonal medication) instead of a vitamin that came in a similar bottle with a similar label. When Solomon and Wallace, assisted by a medical student, inventoried the ward's medical supplies, looking for clues, they found the two bottles sitting near each other on a shelf in a medication cupboard.[3] So the cause of the outbreak was not a malefactor or an infection but a tragic medication error.

Before announcing their discovery, Solomon sought additional proof, hoping to head off speculation that the events on the neonatal and cardiology wards were linked. He learned from Carver, the pediatrics chief, that gastric aspirates from each baby's stomach had been sent to a Ministry of Health laboratory to be tested for bacterial toxins. (A gastric aspirate is a sample of stomach contents taken to recover sputum coughed into the throat and then swallowed.) The medical student assisting Solomon's team retrieved the gastric aspirates from the Ministry and hand-carried them by plane to Irwin Kopin, a toxicologist at the U.S. National Institutes of Health, who confirmed the presence of epinephrine in the babies' stomachs.

Relieved by the quick resolution, Carver—who was himself an EIS graduate—asked the Ministry of Health to invite another epidemiologist from the CDC to investigate the alleged overdoses of digoxin. He was sure that no crime had been committed and believed that an EIS officer would set things straight, just as Solomon had done. The CDC responded by sending one of Solomon's EIS classmates, James W. Buehler, a newly minted pediatrician from California.

Buehler is a soft-spoken, observant, deeply intelligent person with an unassuming manner. As a second-year EIS officer, he already had some field experience, although it was limited. During his first year at EIS, he had tracked a cluster of hepatitis cases to a contaminated well and traced an outbreak of conjunctivitis to a contaminated instrument in an ophthalmologist's office. Now he faced something entirely different.

Buehler arrived at Sick Kids in September 1982, accompanied by his EIS mentor, Clark Heath, a senior scientist known for pioneering studies on environmental clusters of leukemia. Solomon had already filled them in on the general situation. (Buehler and Solomon were both married to EIS officers and lived near each other in Atlanta.) Carver introduced them to Evelyn Wallace, Solomon's former teammate, and to Lesbia Smith, a Puerto-Rican-born physician with a warm manner and a lively sense of humor. Wallace and Smith had been assigned to the case by the Ontario Ministry of Health. As Wallace later testified, she was eager to find an explanation for the increase in mortality that was "not sinister" and "hoped that we might be able to come to some other conclusion, a nice, simple, easy conclusion."[4] After meeting briefly with representatives of the Toronto police, the coroner's office, and several hospital departments, Buehler, Wallace, and Smith set to work.

What could medical detectives do that the Sick Kids doctors and the police had not already done? The doctors had focused on the details of each baby's illness, finding a natural reason for each death. The police, on the other hand, had focused on a particular suspect, seeking legal evidence to build a case against her. The epidemiologists viewed the evidence from a different angle. Unlike the police or the doctors, they looked at all of the deaths at once, as part of a single mission, trying to figure out what all of the cases had in common—somewhat like an FBI analyst examining deaths linked to a single serial killer. However, unlike the police or FBI, they were not concerned with legal issues or with questions about human guilt and motivation, and unlike the hospital staff, they bore no personal responsibility for the babies' welfare. They did not interview the nurses or meet with the victims' parents. Thus, they were emotionally removed from the tiny victims and perhaps better able to analyze the data in a dispassionate way, using graphs and statistics—"people with the tears wiped away" as the EIS saying goes.[5] Another way to say it is that they ignored the horror behind the numbers and plunged on, wherever the data would take them.

Buehler and his teammates employed the same tools and procedures as in any hospital outbreak. Their first challenge was to determine whether the apparent rise could be due to chance, as Carver believed. To resolve this question, they used a standard public health tool called an *epidemiologic curve,* or "epi-curve,"—a line graph that plots deaths or death rates against time—to create a visual representation of the sequence of deaths on the cardiology ward. Because the deaths had taken place more than a year before, the team was able to compare the periods before, during, and after the occurrence of the infant deaths. This is what they found:

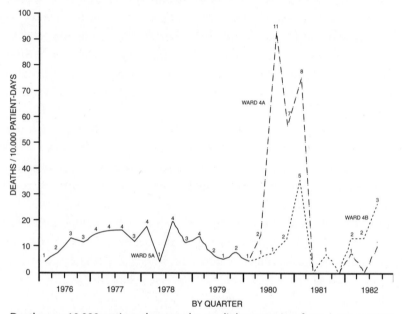

Deaths per 10,000 patient-days on the cardiology service, from January 1976 through September 1982. Before April 1981, the cardiology service was on one ward, Ward 5A; subsequently, it was moved to two adjacent wards, Wards 4A and 4B. Rates are shown separately for each ward. Mortality rates do not include patients who were transferred to the intensive care unit before death. The numbers of deaths are shown above the lines. *Source:* Buehler, J. W., Heath, C. W., Herndon, J. L., Kusiak, R., Smith, L. F., & Wallace, E. M., "Unexplained Deaths in a Children's Hospital," N Engl J Med 323, no. 4 (1985): 211-216.

The epi-curve for the pediatric cardiology wards, originally written in pencil on graph paper, told a simple, striking, and disturbing story. A steep rise in death rates was clearly visible as a large peak that rose above the background level of hospital deaths. The peak covered nine months, beginning in July 1980, three months after the opening of two new cardiology wards (4A and 4B), and ending in March 1981, with the arrest of Susan Nelles. During this time, eight babies had died on Ward 4B and twenty-six babies on Ward 4A, including the four infants Nelles had been accused of killing.

The rise in infant deaths was statistically significant—too large to be due simply to chance. Moreover, no peaks appeared when epi-curves were plotted for the ICU, neonatal ICU, the infant medical wards, or the hospital as a whole. The only rise in death rates was on the pediatric cardiology wards. Also, outside of the nine-month period most infant deaths occurred in the operating room, not on the wards or the ICU.

Buehler and his colleagues were surprised that the hospital had not performed this type of analysis before. They showed the graphs to Carver, who, though quite disturbed, still believed that the rise in infant deaths would be explained away when the epidemiologists looked further.

Buehler, Smith, and Wallace camped out in a basement room at Sick Kids with a huge stack of handwritten medical charts to review the babies' histories, look for statistical patterns, and consider how to proceed. They noted the babies' ages and diagnoses, the time of day at which deaths had occurred, and the time of onset of each baby's deterioration. They noticed that many of the deaths had occurred at night, in the early morning hours, which suggested a recurring problem, such as a medication error, during the hospital's night shift. However, as they poured over the data they realized they could not tell for certain which of the thirty-four deaths were part of the epidemic, because the symptoms of digoxin poisoning—slow or erratic heartbeat, nausea, loss of appetite, and fatigue—can occur for many reasons. How could they look for common factors when they were not sure which deaths to consider? Although they had identified a nine-month *epidemic period*, they still lacked a *case definition*—a clear description of an epidemic case.

To figure out which deaths were likely due to digoxin overdose, Buehler's team asked three medical consultants—a cardiologist, a pathologist, and a pharmacologist—to compare the medical records of the babies who died during the epidemic period with those of babies who died at other times. The cardiologist, Alexander Nadas of Children's

Hospital in Boston, confirmed that the timing of the clinical deterioration of several of the epidemic-period babies was abrupt and unexpected and not characteristic of other babies with the same diagnosis. He scored nineteen deaths as having a clinical pattern that was "consistent" with digoxin overdose and eleven more as "consistent, of special concern."[6] One of the "special concern" deaths had occurred the day before the start of the epidemic period (on June 30), and possibly should have been classified as part of the epidemic. The pathologist, Derek deSa of the Children's Hospital in Winnipeg, reported that autopsy findings were sufficient to explain all but three deaths: two that occurred during the epidemic period and the death that occurred on June 30.

These findings were a second heavy blow to Carver, because Nadas' voice carried weight. Carver himself had recommended Nadas to Buehler, confident that the cardiologist would classify the babies' deaths as expected and natural. Instead, Nadas had categorized almost one-third of the cases as deaths of "special concern." Moreover, the third consultant, pharmacologist Ralph Kauffman of Wayne State University, also scored many deaths as consistent with digoxin overdose. Judging from the progression of the symptoms, Kauffman estimated (at Buehler's request) that a single dose of digoxin, delivered by IV between thirty minutes and four hours before the onset of terminal events (depending on dose and rate of IV infusion) was the most probable and plausible means of drug delivery. Although Kauffman discounted the quantitative data on digoxin levels—because of the small number of specimens and problems with measuring digoxin in preserved or exhumed tissues—he had no innocent explanation for finding digoxin (at any level) in four babies who had not received digoxin as part of therapy.

Based on the consultants' scores, Buehler's team classified eighteen of the thirty-four infant deaths as unexpected, inconsistent with clinical course, and/or suggestive of digoxin toxicity (Category A). These eighteen deaths, plus ten more scored as "consistent" with possible digoxin intoxication (Category B), had occurred between midnight and 6 AM.

While the consultants completed their work, Buehler's team generated hypotheses that might explain the rise in death rates during the nine-month epidemic period, taking into account the ideas and impressions of the doctors and officials they had interviewed when they first arrived. Then they began to test these hypotheses, methodically, one by one:

Was the rise in deaths due to a medication error?

The "accidental poisoning" hypothesis was considered the most likely by the hospital staff. Medication errors like the injection of epineph-rine in place of vitamin E do occur, even in excellent hospitals with excellent staff. The epidemiologists reviewed routines and procedures for administering digoxin, as well as hospital records that monitored drug levels in the blood of children treated with digoxin. However, they found no evidence of a systematic error in the way digoxin was administered during the epidemic period in Ward 4A, Ward 4B, the ICU, or the neonatal ICU.

Perhaps the babies who died during the epidemic period were more likely to have been treated with digoxin than babies who died during other times—which would have made them more likely to be subject to accidental errors. However, the data suggested the reverse: that babies who died during the epidemic period were *less likely* to have been given a dose of digoxin within four hours of their death. Moreover, Buehler's team found no evidence that the babies had been prescribed drugs that might have interacted with digoxin in harmful ways.

The epidemiologic team also considered whether another fluid used in the hospital, such as a liquid medication, IV fluid, or infant formula, could have been accidentally contaminated with digoxin. However, the problem was limited to the cardiology ward, and these fluids were used, night and day, on many different wards. Thus, the most-favored theory—of "multiple, repeated, concentrated, fatal [medication] error" occurring at night in the cardiology wards[7]—did not hold up.

Was the patient population younger and sicker during the epidemic period?

Some doctors at Sick Kids were convinced that the babies on Wards 4A and 4B had been younger and sicker than usual during the epidemic period. A temporary increase in younger and sicker babies could have caused a significant rise in death rates, because babies at risk of death because of congenital heart defects tend to die within the first year of life.

A direct way to test this idea would be to re-calculate the ward-specific death rates used to construct the epi-curve in Figure 1, controlling for age and severity of illness. If the increase in deaths were due to the babies being younger and sicker, the epi-curve peak would flatten out or disappear. However, patients do not spend all of their time in the hospital on one ward, but are moved, as needed, between the medical ward, surgical ward, and the ICU. Because the patients' records provided only an approximate idea of the time spent on each ward, the team did not have reliable data to make the necessary (ward-specific) calculations.

To get at the answer in a different way, the team asked Richard Rowe, Chief of Cardiology at Sick Kids, to evaluate the patient population on the cardiology ward, in terms of age, severity of heart disease (mild, moderate, severe), and prognosis on admission (good, intermediate, poor). Initially, the results of this study suggested that the babies admitted to the cardiology ward during the epidemic period were in fact slightly younger and sicker and had a poorer prognosis for surviving hospitalization—a finding in good accord with Carver's expectations, although the differences were small. Later on, however, it turned out that the team had made a tabulation error that skewed Rowe's results. They had included all of the babies who died during the epidemic period in what was intended to be a random sample of the entire patient population. When the results were recalculated and corrected, no significant difference was found in relation to age, severity of illness, or prognosis.

Two other findings went against the "younger and sicker" hypothesis. First, the condition on admittance of children who died during the epidemic period was less severe than that of patients who died during other periods. In fact, twice as many of the babies who died during the non-epidemic period were in critical status on admission as were the babies who died during the epidemic period. Second, if the patient population was significantly younger and sicker, more deaths would have occurred not only on the wards but also in the operating room and the ICU.

Had something happened at the beginning of the epidemic period that adversely affected patient care?

Buehler's team identified three events involving the medical staff that occurred around the time the epidemic period began, in July 1980. Some interns and residents had participated in a work slowdown in June 1980, as a part of a province-wide labor protest, and a new group of residents joined the cardiology ward a month later, as they do every July. However, the new group rotated out of the cardiology ward six weeks later—one and a half months into the nine-month epidemic period—and the slowdown lasted only two days.[8]

The third event was more disruptive. Three months before the start of the epidemic period (in April 1980) the pediatric cardiology ward was moved to another floor and expanded into two wards, 4A and 4B, that were staffed by two different nursing teams. The number of beds was increased from thirty-eight to forty-two, with twelve beds placed in two "infant rooms" next to the nursing station shared by 4A and 4B. This arrangement helped facilitate round-the-clock nursing care for very young infants who do not sleep through the night like older children. More than half of the babies who died had been cared for in the infant rooms.[9]

The goal of the new arrangements was to improve and enhance patient care. Was it possible that these changes had the opposite effect, adversely affecting the nurses' workloads or some other aspect of patient care? Had the babies in the two infant rooms—which typically contained the sickest infants outside of the ICU—received inadequate care during the epidemic period? Some doctors thought that seriously ill babies had remained on the ward when they should have been transferred to the ICU, which had been especially busy during the epidemic period. They recalled that a hospital committee, aware of increasing distress among the night nurses, had suggested opening an intermediate ICU for pediatric cardiology patients some months before the police investigation began. Another doctor recalled that a cardiologist in Manitoba had referred a higher-than-usual number of surgical patients to Sick Kids, which had caused high ICU occupancy rates.

Buehler's team did detect a gradual rise in surgical patients referred from Manitoba, starting in 1979. However, the number of referrals had already leveled off before the epidemic period began. The team also found that the hospital's optimal ICU occupancy rate had been exceeded in all nine months of the epidemic period, compared with 80 percent of the months during the non-epidemic period. However, they found no recorded complaints from nurses or doctors about understaffing in the cardiology ward or an inability to transfer a sick baby into the ICU when needed. Moreover, several of the babies who died during the epidemic period did not undergo surgery and therefore were less likely to require ICU care.

When the team reviewed the ward's daily nurse-to-patient ratios (which remained fairly constant during the epidemic and non-epidemic periods) they observed that the epidemic period deaths did not tend to occur during days of relative under-staffing or over-staffing. Moreover, an examination of the nursing assignment book indicated that nurses had sometimes been sent off the cardiology ward to help out on other wards, suggesting that the ward had adequate nursing coverage.

The team examined several other factors that might have affected patient care during the epidemic period. However, they found no significant differences between the epidemic and non-epidemic periods in the number of surgical procedures (including complex surgeries), in cardiac catheterizations (used to detect cardiac defects and abnormalities), or the frequency of ward mishaps, as documented by incident reports.

Did the babies who died during the epidemic-period have any special characteristics?

Buehler's team now turned to an examination of the tiny victims. Why had these babies died and not others? Was there a common characteristic that predisposed them to a natural or accidental death, or (though this was never explicitly stated) singled them out as murder victims?

To answer this question, Buehler's team compared the babies who died during the epidemic period with babies who died at other times (the "death comparison study") and with babies who shared their rooms at the time of death (the "roommate study"). They looked at the babies' age, race, sex, place of residence, diagnosis, condition on admission, number and types of surgeries and surgical procedures, medications prescribed, and IV lines. However, no distinguishing factors stood out. The only notable finding—which confirmed their earlier impression— was that a statistically significant proportion of the epidemic period deaths, including all those classified as Category A or B, had occurred at night: 72 percent, as compared with 10 percent of the non-epidemic period deaths.

Was the presence of a staff member associated with the epidemic period deaths?

Having tested a succession of hypotheses and confirmed only a few strong statistical associations—the clustering of deaths at night in wards 4A and 4B—Buehler's team finally decided to consider the disturbing hypothesis favored by the coroner's office and the police: an association between the deaths and a particular person.

To test this hypothesis, the medical detectives employed another standard public health tool—a "risk factor" study—which is designed to identify activities or factors associated with contracting (or dying of) an infectious disease, such as eating a particular food or attending a particular event. In this case, the factor under consideration was "contact with a particular individual"—a factor epidemiologists test for when they suspect a person might be an asymptomatic carrier of an infectious illness. A famous example is Typhoid Mary, a cook who infected more than fifty people in the New York City area with typhoid fever in the early 1900s. A modern example, which occurred not long before the Sick Kids affair, involved a hospital nurse in Michigan who unwittingly infected ten patients over a four-month period in 1980 with streptococcal bacteria, causing life-threatening surgery-site and bloodstream infections.[10] Like Typhoid Mary, the Michigan nurse had no symptoms and no malicious intent. The risk factor study that confirmed her role in the outbreak was not designed to explain how or why she might have caused illness; its goal was simply to establish whether a particular person was in contact with the patients who got ill or died.

Who had worked on the cardiology ward at night during the nine-month epidemic period? As far as the epidemiologists could tell, only doctors and nurses. Hospital support staff, including orderlies, respiratory hygienists, radiology technicians, IV team members, and ward clerks, left the hospital by 10 or 11 PM. Paperwork, garbage, and soiled laundry were picked up before midnight. Many doctors worked on the cardiology ward, especially during the daytime, but their schedules did not coincide with the nine-month epidemic period. The attending physicians rotated among the wards and ICUs on a monthly basis, and the cardiology residents rotated every six weeks. Just in case, Buehler asked for a confidential list of interns and residents who had difficulties of any sort during their training years. But none of those individuals had worked on the cardiology ward for a significant portion of the epidemic period.

The epidemiologists could not rule out the possibility that a stranger had visited the cardiology ward from time to time over a period of nine months. But this seemed unlikely. A hypothetical stranger would have required some medical knowledge and a plausible reason for being on the ward. So the epidemiologic team decided to limit the risk factor study to individuals who had constant around-the-clock responsibility for patient care—which meant the nurses, especially the ten women on the two nursing teams for wards 4A and 4B

Buehler's team created a detailed nursing calendar that specified which nurses were at work each day and night in wards 4A and 4B over the nine-month epidemic period. The hospital's payroll office provided these data in the form of a huge stack of hospital staffing logs. The team initially worked from photocopies, similar to those provided to the police. However, Kathleen Shilton, a hospital nurse assigned to assist them, explained that when the nurses exchanged shifts or helped out on different wards they recorded the changes on the back of each day's log. So the team asked for the originals. With Nurse Shilton's help, they used the handwritten corrected sheets to record on- and off-duty times for each nurse, using half-hour increments. Of the 280 nurses at the hospital, forty-six were on duty at the time of one or more of the babies' deaths, and fifty-seven were on duty within four hours preceding the onset of terminal events.

The epidemiologic team used the data in the nursing calendar to calculate the "relative risk of death" associated with each nurse during the night or day for the twenty-eight Category A and Category B deaths. The "relative risk" was calculated by dividing the death rate during each nurse's on-duty hours (i.e., the number of deaths divided by the number of on-duty hours) by the death rate during her off-duty hours (i.e., the number of deaths divided by the number of off-duty hours). A relative risk of 4, for example, would mean that a baby was four times as likely to die when that person was on duty. A relative risk of 1 would mean that there was no association between the deaths and that person's presence.

Table 1 Associations between Ward Deaths during the Epidemic Period (July 1980 through March 1981) among Patients under One Year of Age and Duty Schedules of Four Members of a Nursing Team.

NURSE	SHIFT	NO OF DEATHS WHILE ON DUTY	NO OF HR ON DUTY	NO. OF DEATHS WHILE OFF DUTY	NO OF HR OFF DUTY	RELATIVE RISK (9.5% CONFIDENCE INTERVAL)*
A	Day	5	677.5	2	2622.5	9.7 (1.5-60.6)
	Night	26	600.0	0	2700.0	Infinity
	Total	31	1277.5	2	5322.5	64.6 (27.2-153.2)
B	Day	1	657.0	6	2643.0	0.7(0.1-4.3)
	Night	21	635 5	5	2664.5	17.6 (6.6-46.7)
	Total	22	1292.5	11	5307.5	8.2 (3.5-19.4)
C	Day	2	694.0	5	2606.0	1.5 (0.2-9 3)
	Night	19	636.0	7	2664.0	11.4 (4.3-30.1)
	Total	21	1330.0	12	5270.0	6.9 (3.0-16.2)
D	Day	0	496.0	7	2804.0	0.0
	Night	18	707.5	8	2592.5	8.2 (3.2-21.0)
	Total	18	1203.5	15	5396.5	5.4 (2.2-13.0)

The "relative risk" of death associated with Nurse A during the night-time was calculated by dividing the death rate during her night-time on-duty hours (26 deaths per 600 hours) by the death rate during her off-duty hours (0 deaths per 2,700 hours). Dividing 26/600 (or any other number) by 0 gives an undefined result that mathematicians–speaking in terms of the "limits" used in calculus–regard as so large that it "approaches infinity."[4] *Source:* Buehler, J. W., Heath, C. W., Herndon, J. L., Kusiak, R., Smith, L. F., & Wallace, E. M., "Unexplained Deaths in a Children's Hospital," N Engl J Med 323, no. 4 (1985): 211-216.

When Buehler's team performed this calculation for the four-member nursing team that included Susan Nelles, this was the result:

For one nurse, the combined relative risk for both the day and the night shifts was 64.6—meaning that a baby was 64.6 times as likely to die when she was on duty as compared with when she was off-duty (Nurse A in Table 1). Moreover, that nurse's relative risk for the night shift was so large that it approached *infinity*—because the death-rate during her off-duty hours was zero (see the figure legend to Table 1). This person was also the only nurse on duty within four hours before the onset of terminal events for all four infants whose bodies tested positive for digoxin even though it had not been prescribed for them. Finally, this person was the only nurse on duty during all four of the cases where Kaufmann had been able to estimate an approximate time of overdose administration.

However, this nurse was not Susan Nelles, whose relative risk for the night shift was 17.2 (Nurse B). The nurse associated with all of the night-time infant deaths was the head of Nelles' nursing team: Phyllis Trayner.

The epidemiologists brought this final result to Carver. They had "peeled back the layers of the onion, one by one," as Buehler says, with the correlation between the deaths and the presence of a particular nurse being the final layer. Each layer had made it more difficult to say that the deaths might be the result of a mistake.

Buehler, Wallace, and Smith wrote up their findings in a confidential report sent first to the CDC for review and then to the Ministry of Health, in February 1983. Buehler returned to Atlanta and waited, but nothing happened. In July 1983, the Ministry of Health submitted the document—known as the "Atlanta Report"—to Attorney General Roy McMurty. But again, nothing happened. The police, having failed to make their first arrest stick, were unable or unwilling to build a case against another person.

Very likely the situation would have been different if the "Typhoid Mary" data had been available before the arrest of Susan Nelles. In that case, the police could have interviewed Nurse Trayner in greater depth, scrutinized her history and background for motive and opportunity, and perhaps elicited a confession. Or, they might have built a strong circumstantial case by translating the epidemiologic data into daily-life terms. They could have explained that Nurse Trayner was on duty for each of the first ten suspicious deaths, which occurred between June and November 1980. Although no deaths occurred while she was on her honeymoon (from August 23 to September 23), a baby died the first night she returned to work (though it was a baby whose death was expected). No deaths occurred in November and December while she worked the day shifts, but (once again) a baby died when she returned to the night shift.[11] This pattern was repeated over and over, with deaths occurring only when Nurse Trayner worked the night shift. Moreover, babies assigned to Nurse Nelles died only on nights when Trayner relieved her during her breaks.

In any case, the refusal by the police to attempt a second indictment, like Judge Vanek's ruling the year before, caused a public uproar. The hospital and the Toronto police were threatened with lawsuits by the babies' families and by Susan Nelles. Attorney General McMurty, hoping to provide a public accounting and restore confidence in the hospital, decided to establish a Royal Commission to determine how the babies had died and to review the police investigation. This Commission, which was called the "Grange Commission" in honor of its presiding jurist, Samuel Grange, became a media sensation that transfixed the Canadian public for a year and a half.

The Grange Commission heard testimony from a long line of witnesses, including nurses, doctors, parents, toxicologists, and hospital administrators. Three months after it began, in January 1984, Buehler was summoned back to Toronto to testify as a member of a panel that also included Wallace, Smith, and Robert Kusiak, a government statistician who helped design their studies and analyze the data. Buehler recalls that the testimony seemed to take an eternity! On the first day, the Council for the Crown—a stout, humorous person with a plummy accent whom a reporter likened to Charles Laughton in *Witness for the Prosecution*[12]—led the panel through a direct examination of their investigative activities, from arrival at the hospital to the submission of the Atlanta Report. Along the way, he asked Buehler to define epidemiology, explain the role of the CDC, and describe the tools and uses of public health statistics. About half way through, he referred to a critique of the Atlanta Report written by R. Brian Haynes and D. Wayne Taylor, senior scientists at McMaster University, at the request of the Hospital for Sick Children. The so-called "Haynes Report," which Buehler had received only a few days before, was cited repeatedly during the cross-examination, which lasted three days.

The public health panelists were questioned by a succession of lawyers representing the hospital, the Toronto police and coroner's office, the families of the victims, the Registered Nurses Association of Ontario, and members of the Sick Kids staff, including Nelles and Trayner. The lawyers focused on issues of potential interest to their clients, some of whom were filing lawsuits or hoping to defend themselves against lawsuits. The attorneys for the doctors and the hospital, for example, emphasized that Sick Kids had taken early action to increase physician coverage on the night shift and open an intermediate-care ICU. The attorneys for the babies' families wanted to know whether their children's deaths had been scored by Nadas or Kauffman as consistent with digoxin overdose.

Depending on their agendas, the lawyers either treated the epidemiologists with great deference—"like heroes who walked on water"—or made them out to be incompetent fools. The lawyer for a nurse on Trayner's team, for example, rattled off a list of errors in the Atlanta Report, including two transposition errors, a missing date on the Report's transmittal letter, the tabulation error in the analysis of Rowe's data (detected by Haynes), and the failure to include the hospital's chief pathologist on the list of people the team had interviewed.[13] On the other hand, the attorney representing the Attorney General and the coroner's office pointed out that—despite critiquing its study designs, quality of data, and statistical methods—the Haynes Report agreed with each major conclusion of the Atlanta report: that the increase in deaths was real and large; that the deaths were clustered among infants on the cardiology ward between midnight and 6 AM; and that it was "not possible on the data available that any person other than Nurse Trayner could have been more strongly associated" with the most suspicious deaths, because "she was associated with all of them."[14]

As the cross-examinations continued, it became increasingly clear how difficult it would be to use the epidemiologic data in an actual murder trial. The attorney representing the Registered Nurse Association and thirty-nine individual nurses, for example, poked legal holes in the data suggesting that only nurses—as opposed to doctors, strangers, nurse supervisors, and personnel from other floors—had access to the cardiology wards late at night.[15] Moreover, Trayner's lawyer, who conducted the final cross-examination, implied that the epidemiologists' judgments had been influenced by publicity surrounding the arrest of Nelles, by the testimony at her preliminary hearing, and by statements Trayner had given to the police (which the panelists had not in fact read).[16] Trayner's attorney concluded his argument by suggesting that the person who harmed the babies could have timed his or her actions to frame Trayner and her nursing team.[17]

As he answered the questions as best he could, Buehler tried to ensure that the lawyers, whether trying to make him look good or bad, did not over-interpret, under-interpret, or misuse the epidemiologic data. He reminded them that epidemiologists are not criminal investigators and that their aim is to detect statistical associations, based on the data at hand. While each lawyer represented the interests of a particular person or group, the panelists were there to present their statistical findings, as precisely and accurately as they could.

The cross-examination that Buehler recalls most vividly was conducted by Ian J. Roland, an attorney representing Sick Kids. Buehler was dismayed that the same hospital officials who had welcomed him to Toronto were now trying to discredit the Atlanta Report, which he considered a joint effort of the Ministry of Health, the CDC, and the hospital itself. As a junior person, he was also worried about holding his own with the lawyers and scientific experts the hospital had hired to pick apart his studies and undermine his conclusions. He took heart from the realization that he and his colleagues were thoroughly familiar with the data, which they had gathered and analyzed themselves, and that their credibility came from their depth of knowledge. He was also encouraged when Haynes—who attended the hearing to assist attorney Roland—told him he was doing a great job when they crossed paths in the hallway.

Roland began with a barrage of detailed questions about the methods and study designs described in the Atlanta Report. In response, Buehler explained the choices the team had made when lack of data prevented them from posing questions in the most direct way or from performing additional statistical tests. He defended their reliance on the subjective assessments and scoring systems devised by Nadas, Kauffman, and Rowe, and attested to the accuracy of the process used to abstract information from the babies' medical charts.

By the time Roland turned his attention to the study that implicated Nurse Trayner, Buehler was focusing as hard as he could, thinking through every question as carefully as possible, trying not to fall into any traps. The day before, the lawyer for the Registered Nurses Association had questioned the use of the hospital payroll sheets to construct the nursing calendar. Buehler described what his team had done to ensure the accuracy of the data but admitted that "it is possible the information was not 100 percent complete."[18] Roland tried to build on that admission to cast further doubt on the study. After implying it was a mistake to rely on a Sick Kids nurse to build the calendar, he waved a sheet of paper—apparently a nursing timesheet—in front of Buehler, and demanded to know if it was a source of data available to Nurse Shilton. With a dramatic flourish of his own, Buehler reached out and flipped the paper over, so everyone could see that the back of the page was blank. "These are Xeroxes," he said, "and Nurse Shilton used originals" that included handwritten changes made at the last minute.[19] Roland backed off, addressed a few more questions to Smith and Kusiak, and ended his cross-examination not long after.

The Grange Commission continued for eight more months, ending in September 1984. In an official report, issued in January 1985, Justice Grange concluded that at least eight babies and possibly as many as twenty-three had died of overdoses of digoxin. However, he also found that there was insufficient evidence to charge anyone with the crime. He also ruled the police and the crown attorneys had

acted in good faith. Finally, he agreed with Judge Vanek that Susan
Nelles was innocent and recommended that she be compensated for
legal costs.[20] Thus, the Grange Commission ended on an unsatisfactory
note, with relief that it was over, but doubt that justice had been done.

No one wants to believe that anyone, let alone a nurse, would deliberately inject poison into a baby's IV bag. So it is not surprising that over
the years many individuals, including both scientists and amateur sleuths,
have tried to prove that the deaths were either natural or accidental. One
early theory, raised during the Grange Commission, is that the digoxin
readings were a measurement artifact caused by cross-reaction with naturally occurring chemicals.[21] A second "natural deaths" theory is that the
autopsies suggesting foul play were performed by a nefarious pathologist named Charles Smith who lost his license in 2011 for professional
misconduct and poor quality work on suspected cases of "shaken baby
syndrome" and other suspicious child deaths.[22] Because Smith began conducting autopsies at Sick Kids in 1981, it is possible that he performed
autopsies on babies who died toward the end of the epidemic period.
However, his name is not mentioned in contemporary accounts.[23]

Be that as it may, neither theory accounts for two major findings
in the Atlanta Report: that a real and large increase in infant deaths
occurred on the cardiology ward over a nine-month period and
that most of the deaths took place between midnight and 6 AM.
Discrediting the digoxin data or the autopsy data, or both, does not
make the rise in deaths disappear or explain why the deaths occurred at
night. A similar objection holds for a third, more recent theory, which
suggests that the babies were accidentally poisoned by a toxin found
in natural rubber used to manufacture disposable plastic syringes and
intravenous devices.[24] Why would this toxin kill infants only on the
cardiology ward of Sick Kids, and not in the ICU, the neonatal ICU,
or the medical wards of Sick Kids or other hospitals? And why would
the babies have died only at night?

This question remains a major sticking point for theories that try
to explain away a "sinister" explanation. As Wallace said when the
Nursing Association lawyer asked why Buehler's team had discounted
the idea that infants at other hospitals might also be dying from an

undiagnosed condition: "We know of no disease condition which would predetermine death between the hours of midnight and six."[25]

Psychologists and legal scholars who accept the possibility that a nurse could have killed her own patients have tried to figure out what could possibly motivate such a crime. A study of 90 criminal prosecutions of alleged healthcare serial killers suggests that nurses who induce cardiac arrest in their patients might suffer from "a professional version of Munchausen Syndrome by Proxy," a psychiatric disorder in which a "primary care giver induces a health crisis in his/her child for the purposes of getting medical attention."[26, 27] Another possibility is that these individuals are like firefighters who set fires to generate excitement and create opportunities to perform heroic rescues.[28] Whether these ideas apply to the Sick Kids case, we do not know.

Over the years Buehler sometimes hoped for a deathbed confession that would bring resolution and a measure of closure. However, that hope ended in 2011, when Phyllis Trayner died of cancer, in her mid-fifties, never arrested and never exonerated. She had raised two children as a divorced single mother, working as an occupational nurse at a large corporation where she took care of adult patients only. She maintained her innocence until the end.[29]

Susan Nelles, who was on her honeymoon in Banff when the Grange Commission report was released, went on to a distinguished career in nursing and nursing education. Her experience of being falsely accused led her to support a less traditional and more assertive role for nurses. She urged her colleagues to speak out and demand answers whenever they suspected something might be wrong. In 1991, she used her settlement with the Ontario Government to endow a Queen's University nursing scholarship in memory of her brother, Dr. David Nelles, and her father, Dr. James Nelles, who had died during the preliminary hearing. In 1999 she was awarded an honorary doctorate in law from Queen's University, her alma mater, for her work in promoting integrity in nursing.

Today, James Buehler is a Professor of Health Management and Policy at the Drexel University School of Public Heath in Philadelphia, where he moved after retiring from CDC in 2013. Looking back, he regrets that Phyllis Trayner neither confessed nor had her day in court.

She was, as he says, tried in a court of opinion only. Nevertheless, he has the satisfaction of knowing that his work at Sick Kids helped improve patient safety in North America and beyond. In a 1984 paper published in the *New England Journal of Medicine*,[6] he and his teammates made two recommendations that were widely adopted by the hospital community. The first recommendation is that hospitals should dispense drugs in single doses from a central pharmacy instead of keeping bottles of medicine on open shelves in each ward. The second recommendation is that hospitals should institute routine mortality surveillance—tracking of deaths that occur in the hospital—by keeping detailed records on all deaths and by analyzing mortality data to look for unusual patterns. As part of these efforts, hospitals should record not only the cause of death, but also the time of day when the patient died, where in the hospital the patient died, and which staff members were on duty. Hospitals should also calculate ward-specific death-rates over time, as Buehler's team did in creating the epi-curve that identified the nine-month epidemic period.

Although public health experts had been advocating these quality improvements for some time, the experience at Sick Kids demonstrated their importance in a dramatic and unequivocal way. If single-dose dispensing had been in place at Sick Kids in 1980, it could have limited unauthorized access to digoxin and also prevented the medication error detected by Solomon and Wallace. And if mortality surveillance had been in place, the hospital might have detected the rising death-rate on the cardiology ward, stopped the epidemic at an early stage, and provided the police with solid evidence that might have led to a conviction. Today, single-dose dispensing and routine mortality surveillance are accepted practice at most hospitals in Canada, the United States, and many other countries.

4

Obsession or Inspiration

On Monday, August 2, in the bicentennial year of 1976, an official of the Pennsylvania chapter of the American Legion reported something strange: same-day obituary notices for four middle-aged Legionnaires, briefly sick with an influenza-like illness, who had attended a convention in Philadelphia during the third week in July. Could this be a coincidence, or had they all died of the same thing? Had a few conventioneers shared a contaminated meal or drank from a contaminated source? Or—even worse—could they be the first known victims of the dreaded "swine flu," a highly dangerous disease whose emergence had been predicted eight months before?

Back in January an eighteen-year-old soldier in a boot camp in Fort Dix, New Jersey, unwilling to limit his training activities while suffering from the flu, had died of respiratory failure after an all-night march. Using the laboratory methods available in 1976—lacking the modern molecular tools that quickly decoded the genes of the H1N1 flu virus detected in 2009—scientists determined that the surface proteins of the virus that killed the soldier were similar to those of a swine virus thought to be related to the "Spanish Flu"—the disease that killed ten million people in 1918–19. Fearing the next worldwide epidemic (or "pandemic") of influenza, the U.S. Center for Disease Control (CDC)[1] made the controversial and expensive decision to vaccinate the entire U.S. population against the Fort Dix virus, believing that in the face of extreme danger and uncertainty it is wise to err on the side of caution.

As the winter flu season passed, however, and no pandemic emerged, public concern about swine flu—never great to begin with—began to wane, and it looked as though political, financial, and liability issues would derail the vaccination program, despite the backing of President Gerald Ford. But then, to everyone's surprise, in the weeks following the Bicentennial celebrations in July, the scary predictions were suddenly and horribly fulfilled, when more than 150 people were hospitalized with an acute respiratory disease, and twenty-two of them died. By the end of the summer, the tally had risen to 182 cases, with twenty-nine deaths. Was this swine flu after all?

The U.S. public, bombarded by daily news stories, was disturbed and frightened by the outbreak in Philadelphia, even though swine flu was quickly ruled out as a possible cause. Laboratory tests for influenza were negative, there were hardly any runny noses or sore throats, and the illness did not spread to family members, as would be expected for flu. But that did not seem to matter. The public was primed and ready to believe that something big and scary was about to arrive, and it had. Looking back today, what happened in July 1976 was in some ways similar to what happened twenty-five years later, in October 2001, when the anthrax incidents caused hysteria among a public already traumatized by September 11.

Most infectious disease mysteries are quickly solved by testing patient samples in the laboratory and by identifying "routes of transmission"—the ways in which a microbe or toxin is spread (e.g., via food, water, or animals, or person-to-person by touching, coughing, or sneezing). But this time, despite a huge public health investigation involving hundreds of on-the-ground ("shoe-leather") epidemiologists and laboratory experts, no solution emerged for months and months. Some said it was a Communist Plot or a terrorist attack. Others thought that the cause might never be known.

The Epidemiologic Investigation

The symptoms of the new disease included fever, chills, headaches, muscle aches, and a dry or sputum-producing cough. Although most patients recovered within a few weeks, some became seriously ill, developing high fever and severe pneumonia (infection of the lung) that was sometimes fatal. Respiratory illness that can lead to pneumonia—the sixth most common cause of death in the United States—may be due to many types of human pathogens, including bacteria, viruses, fungi, and parasites. It can also be caused by biological or industrial toxins. The job of the public health investigators was to figure out which of these agents was causing the disease and to advise doctors, patients, and the public on what could be done to prevent further spread.

At first things moved quickly. The Pennsylvania health department identified twelve more suspect cases within the next few days and began sending blood samples and autopsy tissues to the CDC in Atlanta for testing. At the request of the State Health Commissioner, the CDC also assembled an "Epi-Aid" team of EIS officers—trainees in the Epidemic Intelligence Service, the national training program for disease detectives—who were sent to Harrisburg, Pittsburgh, and Philadelphia on August 2, and were assigned the task of visiting patients throughout the state. By August 4, CDC laboratories had ruled out all types of influenza (including swine flu) and a foodborne microbe that can cause respiratory symptoms (*Streptococcus* group A bacteria). The laboratories also ruled out the most common bacterial cause of pneumonia—*Streptococcus pneumoniae*—as well as infections caused by *Staphylococcus aureus,* which can sometimes lead to pneumonias in infants, young children, and patients recovering from influenza. On August 5, the Governor of Pennsylvania and the director of the CDC formally announced that the illnesses were not caused by influenza virus or by intentionally-spread plague bacteria—a special concern during the Cold War years. On August 6, with no microbe identified, the CDC sent a second EIS team to Philadelphia to look for evidence of a disease-causing chemical or poison. On the same day, the CDC director testified at a Senate hearing that the number of cases was already starting to diminish and that the disease was not contagious.

Despite this encouraging news, the jolt of fear generated by the outbreak convinced a previously skeptical Congress to support the swine flu vaccination campaign proposed by the CDC. Several lawmakers withdrew their objections to President Ford's plan for indemnifying manufacturers who agreed to include the Fort Dix swine flu virus as one of three inactivated viruses in the vaccine for the 1976–77 flu season. So the National Swine Flu Immunization Program of 1976 went forward after all.

The first step for the epidemiologists in Pennsylvania—who were led by Leonard Bachman, the state secretary of health, William Parkin, the state epidemiologist, and David Fraser, the leader of the CDC team—was to determine how the causative agent was transmitted. Following up on the apparent connection with the American Legion Convention—which hosted four thousand veterans from all over the state of Pennsylvania—the epidemiologists designed a phone survey of people who had stayed at Philadelphia hotels during the convention (i.e., between July 21 and July 24). The survey, carried out by the Philadelphia police force, revealed that the hotel guests who fell ill with fever had stayed at one of two hotels that had hosted Legionnaires. One of them was the Bellevue-Stratford Hotel on Broad Street, a grand and venerable institution—the Philadelphia equivalent of New York's Plaza Hotel—which had served as the convention's headquarters. All of the conventioneers, including those who slept elsewhere, had attended meetings and social events at the Bellevue-Stratford, which also hosted cocktail parties for candidates for American Legion offices. The time between arriving at one of the two hotels and falling ill was two to ten days. The rate of illness among the Legionnaires was 6 to 8 percent.

The epidemiologic link between the patients and the hotels strengthened the epidemiologists' presumption that the deaths from respiratory failure were related—part of a single outbreak. It suggested that the case might have a common origin—a "pin-point" source— such as a particular food served at a hotel restaurant or a particular activity, such as swimming at a hotel pool. A daily survey of the city's hospitals, carried out by public health nurses, also confirmed an association between the Bellevue-Stratford and the outbreak. Although the nurses identified thirty-eight pneumonia patients who had not been hotel guests, nearly all of them reported walking down Broad Street between July 21 and 24; they had not entered the Bellevue-Stratford but had passed in front of it. The epidemiologists referred to these patients as cases of "Broad Street pneumonia," a name that reminded them reassuringly of John Snow, the "father of epidemiology," who removed the handle of the Broad Street pump in London in 1854 to stop the waterborne spread of cholera. Each new case identified by the nurses was recorded on a master list of cases maintained at the Pennsylvania health department in Harrisburg. A review of emergency room records conducted by EIS officers indicated that no other cluster of pneumonia cases had occurred in Philadelphia over the summer, aside from the outbreak among Legionnaires and Broad Street pedestrians.

So far, so good. Based on the survey findings, the Pennsylvania Department of Health refined its definition for identifying additional patients. A case of convention-associated disease or Broad Street pneumonia was defined as a person with a temperature of 102 °F or higher and a cough—or a fever of any degree with radiographic evidence of pneumonia—in anyone who had a physical association with the American Legion Convention in Philadelphia in July. The EIS officers visited each patient, whether identified through the hospital survey or through a physician referral, reviewing medical charts, speaking with doctors and family members, and collecting clinical specimens. Working from a lengthy list of questions, the EIS officers recorded each patient's age, sex, date of illness onset, list of symptoms, drug treatment, travel history, and previous medical history. They also asked detailed questions about the patient's activities while in Philadelphia. The EIS officers were instructed to report back to headquarters—the Pennsylvania state health department in Harrisburg—after each interview.

A Baptism of Fire

Entrusting trainees with big jobs and major responsibility is a key practice of the EIS program, which introduces students to the challenges, excitements, and dangers of a career in public health. James Marks—one of the EIS trainees—remembers the 1976 investigation in Philadelphia as a wonderful but challenging experience, a "baptism of fire." Dedicated, eager, and (as he says) completely green, Marks had just joined EIS, completing a month-long orientation in Atlanta and then moving to Columbus with his wife and infant daughter for a two-year assignment at the Ohio health department. On his first day of work in Ohio—Monday, August 2—the EIS office in Atlanta asked him to fly out to Pittsburgh to help investigate cases of respiratory disease that might be swine flu.

Arriving at the Allegheny County health department the next morning, Marks and two other EIS officers were handed a long list of patients to visit in Western Pennsylvania. Over the phone, the team leaders in Harrisburg dictated a series of questions to ask each patient (or the patient's friends or family members, if the patient was too ill to speak). The EIS officers took down the information by hand and made copies on a mimeograph machine, a process that took up most of the morning. They left the health department around noon, followed by a cameraman from a local TV station who accompanied them as far as Arby's Restaurant. Over a lunch of hero sandwiches and coffee, they divvied up the list of patients, deciding who would go where. Marks rented a car to go from town to town in the area south and east of Pittsburgh, while his colleagues visited hospitals in Pittsburgh and its northern suburbs. Because the patients were hospitalized in many towns in many different hospitals—usually with only one patient per hospital—it would take Marks a full three days to interview all the patients on his list.

At the first hospital, Marks was besieged by the members of the hospital's infection control committee, who wanted advice on what to do if the Legionnaire in their care turned out to be the first local case of swine flu. Should they close the hospital? Should they set up a special section of the emergency room to receive other swine flu patients? What treatment should they provide and what infection control precautions should they take?

Marks—who had yet to see a single case of the mysterious disease and had completed his pediatrics residency only six weeks before—realized that he was in over his head, with little more knowledge than the hospital staff. But Marks is a resourceful person, with a warm smile and a positive presence, an ideal demeanor for someone whose job requires eliciting information from people under stress. Recalling the adage that "in the land of the blind, the one-eyed man is king," he helped them organize their thoughts on preparing for a pandemic, assuring them that they would know more soon. In regard to treatment, he told them that some patients at other hospitals were on antibiotics (in case it was a bacterial pneumonia) or steroids (to reduce inflammation and help them breathe). All were receiving supportive care. In regard to infection control, he agreed that it was good practice to implement precautions appropriate for flu, such as isolating the patient in a single room and wearing a surgical mask while providing care.

The interview with the sick Legionnaire went smoothly, because the patient was well enough to speak for himself and was eager to help. A family member stood by, gloved, gowned, and masked, while Marks asked where the patient had stayed in Philadelphia, which convention events he had attended, and when he had first felt ill. The patient told him that the doctors and nurses were nervous about taking care of him, and that one or two staff members had refused to enter his room. Marks was nervous too—especially with a new child at home—but he had been in similar situations before, during his internship and residency. In each of those situations he had observed standard infection control precautions, and (he reminded himself) he had never gotten infected.

When the interview was over, Marks called Harrisburg on a pay phone to report the results. However, he could not get through. He tried again several times before going on to the next hospital, in another small town outside of Pittsburgh. By now, Marks understood that he needed to be better prepared. He pulled into a gas station and bought all available newspapers, including a Pittsburgh paper, a Philadelphia paper, and the town's local paper. He read everything he could find about the epidemic before proceeding. The newspapers called the illness "Legionnaires' disease" (LD)—a popular name that would later become official.

At the second hospital, Marks spoke with worried doctors and nurses who knew how devastating a flu epidemic could be. Then he began his second interview, which was more difficult than the first, because the patient was quite ill and unable to answer all of the questions. Once again, Marks tried to call in his results, but the line was still busy.

Marks continued visiting patients all day and tried (and failed) to call in after each interview. At supper time, the line was still busy. Back in the hotel, he watched the report on the epidemic on the eleven o'clock news—there was only network news in 1976; no 24/7 cable news channels—which helped him formulate his questions to headquarters when he finally got through, after midnight. (It turned out that non-stop calls from the press had tied up the phone lines all day!) Feedback from headquarters included snippets of information provided by other EIS interviewers. Some patients had eaten at Bookbinders (the oldest seafood restaurant in Philadelphia) or at Horn & Hardart's (the first automat in the United States). And many patients had stayed at the Bellevue-Stratford. Marks, who grew up in Buffalo, was not familiar with these Philadelphia landmarks.

On the second and third days Marks continued to read the newspapers in each town he visited. He no longer wasted time trying to get through to headquarters if the line was busy on the first try (which it always was), but reported in late at night. By the third day, it was clear that the local hospitals were not experiencing a second wave of admissions involving the patients' families and friends, as would be expected for flu—an observation that jibed with the findings of the CDC laboratory. Moreover, the hospitals' laboratories had already ruled out common bacterial pneumonias, which do not spread as quickly as flu. The Pennsylvania papers quoted local experts such as Cyril Wecht, the medical examiner of Allegheny County, who speculated that LD might be caused by a chemical or toxin.

On August 5, Marks visited the Pennsylvania health department in Harrisburg, where the details from the interviews were inscribed on a large wall-sized chart to help identify similarities—disease "risk factors"—that might lead to a solution. But no additional clues had emerged, although the connection with the Bellevue-Stratford was strongly confirmed. Marks spent the following day in Philadelphia, helping out as needed, before returning to Ohio, now fully invested in his new career as a public health investigator.

Meanwhile, puzzled but hopeful, the Philadelphia team carried on. The team leaders—Bachman, Parkin, and Fraser—decided to employ two additional investigative tools: a risk-factor questionnaire and a case-control study. The questionnaire was administered to thousands of Legionnaires who had attended the convention and returned home. Its aim was to figure out whether any activities (meals, drinks, meetings, parties), environments (hotel rooms, lounges, hospitality suites), or occupations (housekeeper, waiter, check-in clerk) were associated with illness, whether mild or severe. Because the epidemiologists did not know what risk factors to look for, the questionnaire was quite long, covering many possibilities.

Lacking a master list of people who attended the convention— or an accurate head count—the American Legion arranged for its Pennsylvania posts to distribute and collect the completed questionnaires. Legionnaires' wives, who were invited to functions at the Bellevue-Stratford but slept apart from their husbands at the Ben Franklin Hotel, were included in the survey. An analysis of the results confirmed what they already knew: illness was associated with spending time in the lobby of the Bellevue-Stratford or on the sidewalk in front of the hotel between July 21 and 24. Eleven of the victims had spent only one day (July 23) at the convention.

The case-control study involved an intensive comparison of forty sick conventioneers with forty local people who did not fall ill (the healthy controls, matched by age and sex) and with forty patients who had experienced other types of pneumonia. Throat swabs, blood samples, and stools were provided by surviving case-patients and controls, and lung biopsies were obtained from patients who had died. All of the samples were sent to the Pennsylvania State Laboratory and to the CDC. (By the time the case-control results were collated, fifteen more people had died.) The case-control data indicated that people who were most likely to acquire LD or die from it were older men who were cigarette smokers or who suffered from chronic lung disease. However, this information did not help very much, because the conventioneers were all men, and cigarette smoking and chronic disease are risk factors for all types of pneumonia, no matter the cause.

While all this was going on, the CDC environmental team was taking surface swabs from all over the Bellevue-Stratford Hotel—from the carpeting, the wallpaper, the kitchen appliances, and the furniture in the rooms where hotel guests ate and slept. (Back in Atlanta, the swabs were tested, without success, for heavy metals, pesticides, and other toxic substances. The swabs were also placed in Petri dishes to see if any microbes would grow, but none did.) The members of the environmental team collected samples of everything they could find in and around the hotel, including dead birds and rodents and pigeon droppings on the roof. They kept generating and testing new hypotheses. They would check out each new idea (e.g., contaminated elevators or lobby bathrooms), taking new environmental samples and asking new questions about where LD patients had spent their time. But none of their ideas panned out.

The environmental team paid special attention to the hotel's air conditioning system, whose roof-top cooling tower was old and leaky. It was an old-style, central air conditioning system that worked by circulating water from the cooling tower through multiple cooling coils distributed throughout the building (similar in concept to a water-heated radiator). According to the questionnaire, few of the LD patients (and none of the Broad Street patients) drank hotel tap water, so the team was not surprised when nothing grew out of water samples taken from the hotel's pipes. But the air conditioning system seemed a possible source of airborne microbes, because the unit in the lobby was right over the registration desk, and the unit on the roof could have sprayed aerosolized water on the street below, accounting for the Broad Street cases. Unfortunately, however, the filters in the lobby air conditioning unit had been changed immediately after the convention ended, and the new filters had been coated with motor oil to help them catch dust. Nothing grew out of surface swabs taken from the lobby or roof units or from the filters or the water in the cooling towers.

All of the CDC staff members assigned to Philadelphia, including the EIS officers, were housed in the Bellevue-Stratford, which had started to empty. But none of them became ill. With the exception of one air conditioning maintenance worker, the hotel employees did not fall ill either.

The Laboratory Investigation

The CDC teams sent a constant stream of patient specimens, autopsy tissues, and environmental swabs to the CDC laboratories in Atlanta.

With all the most obvious causes of pneumonia ruled out, the CDC laboratory decided to check less common pneumonia-causing microbes, whether viral, bacterial, fungal, or parasitic. Over the next weeks and months, they tested for more than twenty-six microbes, using a variety of in vitro and in vivo methods.[2] All were negative.

Most bacteria can be grown in twenty-four to forty-eight hours on special media or in laboratory animals (guinea pigs or mice) and identified by the size and shape of their colonies or by the size and shape of individual bacteria under the light microscope. The scientists ran tests for bacteria that cause whooping cough, tularemia ("rabbit fever"), typhoid fever, and "walking pneumonia" (caused by *Chlamydia pneumoniae*). But no bacteria grew. Viruses may be viewed under an electron microscope, or they can be detected indirectly, by detecting virus-specific antibodies in blood serum from patients who have recovered. But no viruses—mumps virus, measles virus, choriomeningitis virus, or adenoviruses—were found by either method. Additional techniques for growing viruses in the laboratory—in guinea pigs, mice, chicken eggs, monkey cells, and human cells—did not work either. Tests for three parasitic diseases, fifteen species of yeast, and two species of mycoplasma (a special type of bacteria that lacks a cell wall) were also negative.

As time went on, the frustrated biologists tested additional microbes that only rarely affect the lungs. They also checked out special pathogens—highly dangerous viruses never seen in the United States, such as Marburg, Ebola, and Lassa hemorrhagic fever viruses. Still nothing. The CDC toxicology laboratory and a laboratory at the Georgia Institute of Technology performed radioactive assays on autopsy tissues, looking for heavy metals such as cadmium, chromium, mercury, arsenic, thallium, nickel, or cobalt. They also used gas chromatography and mass spectroscopy to separate out the components in each sample, looking for unusual spikes that might indicate the presence of a contaminating substance. But there were no spikes. The scientists' initial confidence turned to worry and discouragement.

The months rolled by, and there was no solution. The association between the patients and the Bellevue-Stratford Hotel remained the only solid clue. Working in an unaccustomed glare of media scrutiny, the EIS officers and their Pennsylvania colleagues continued to follow every lead and rumor, no matter how slim. They investigated two unusual cases of pneumonia in residents of New Jersey (ruled out as LD) and several more in travelers who had attended an International Eucharistic Congress at the Bellevue-Stratford in early July (ruled in). When the local "pigeon lady" who spent her days feeding birds on Broad Street was hospitalized with fever, they wondered if might be psittacosis pneumonia, a bacterial disease transmitted by birds. But the laboratory test for the bacteria that causes pstittacosis (*Chlamydia psittaci*) was negative. They interviewed the family of an eighty-two-year-old woman who died a few days after using the ladies room at the Bellevue-Stratford, and they spoke with an airplane pilot who spent six hours sleeping in the hotel and fell ill after landing his plane in another city. Neither interview turned up any new clues. They tracked down a magician who entertained children at the hotel to find out if he used any unusual devices or chemicals in his act. (He did not.)

Reporters accompanied the EIS officers everywhere they went in Philadelphia, just as they had in Pittsburgh, on Marks' first day in town—except into the hospital rooms of LD patients, where they were afraid to go—until the Philadelphia health department set limits by restricting press access to twice-daily briefings. The tabloids published a constant stream of colorful theories about the accidental release of biologic or chemical weapons by the U.S. Army (officially denied by the Department of Defense on August 30) and about terrorist plots linked to the Bicentennial celebration or to upcoming visits from dignitaries such as the Pope, who cancelled his planned stay at the Bellevue-Stratford. One story said that a Soviet agent had opened his attaché case in front of the hotel, releasing an undetectable airborne microbe. Another said that cigarettes distributed to the conventioneers in souvenir packages (in pre-cancer-warning days) were intentionally contaminated with bacteria. The FBI was concerned enough about the Pope's visit and a possible campaign stop by President Ford that they consulted the CDC about safety precautions, as did the CIA.

In mid August, a more plausible theory was suggested by a pathologist at the University of Connecticut who detected high levels of nickel carbonyl in lung autopsy samples from LD patients. At first blush, this explained everything. Nickel carbonyl is a highly hazardous industrial compound—nicknamed "liquid death"—known to cause lung damage in people who work in nickel refineries and nickel-processing plants. It is used in the manufacture of Plexiglas, magnetic tapes, latex paints, and other products. Inhalation of nickel carbonyl can cause LD-like symptoms (headache, chest pain, coughing, fatigue, and pneumonia)—though not fever—has an incubation period of a few days, and can be inhaled or absorbed through the skin.

But the nickel carbonyl theory had a serious flaw (in addition to not causing high fevers). How could this chemical persist undetected for days in sufficient concentration to infect more than 150 people? There was no accidental source of the chemical near by, and no evidence of terrorism or sabotage. In short, investigators found no evidence at all to back it up.

Airborne diseases have always inspired puzzlement and imaginative theorizing. In view of the Bicentennial, it is interesting to note that an airborne disease was also a major concern to Philadelphians in July 1776, when the Continental Congress issued the Declaration of Independence. Although that disease was a familiar one—yellow fever—which appeared nearly every summer, its origin was an endless source of speculation. Some physicians thought it came from "bad air" (which was also thought to be the source of malaria), or from fumes from blocked sewers or from rotten cargo on the waterfront. Abigail Adams speculated that it might be caused by "a putrid state of air occasioned by a collection of filth, heat, and moisture."[3] The mystery was solved 125 years later, when Walter Reed, of the U.S. Army, demonstrated that yellow fever (like malaria) is spread by mosquitoes.

The Laboratory Investigator

At the end of August, after three and a half weeks of field work, the CDC teams, stymied, returned home from Philadelphia. They had identified 221 cases of LD (including six patients who attended the International

Eucharistic Congress at the Bellevue-Stratford in early July), and thirty-four people had died. All known causes of pneumonia had been ruled out. Although the outbreak had already stopped, with no additional cases identified after August 18, public worry continued unabated.

Enter Joseph McDade, a young scientist who began as a bit player in the overall drama. McDade was not an epidemiologist, though he used epidemiologic data in his laboratory studies. He was a microbiologist, new to the CDC, and quite junior. His boss, a famous microbiologist named Charles Shepard—a pioneer in the study of leprosy—had asked him to rule out one particular infection, Q fever, as the cause of the Philadelphia outbreak, as part of the general effort to leave no scientific stone unturned. Q fever is a zoonotic (animal-borne) respiratory disease, first described in Australia in the 1930s, that can sometimes lead to pneumonia. The "Q" stands for "Query"—a reminder that Q fever was the mystery disease of its day.

Source: CDC Public Health Image Library.

Joseph E. McDade, using a transmission light microscope, 1977.

McDade is a lean, clean-cut man, good-looking in a quiet, bespectacled, scientist-like way. He combines the studious, inward look of an academic with the intensity and determination of an athlete. One can imagine him turning a double play at second base or making a perfect entry pass into the low post. But he would be motivated by doing his personal best rather than by competing with others. Like Charles Shepard,[4] he is a low-key, meticulous scientist who expects to get to the bottom of things through well-designed and carefully executed experiments. He has total faith that scientific rigor and patience lead to knowledge.

McDade was given the task of testing for Q fever because he is an expert on rickettsia, a type of bacteria that lives inside other cells and causes such diseases as typhus and Rocky Mountain spotted fever. During the 1970s, rickettsial experts were in low demand and low supply. In fact, McDade was the only rickettsial expert at the CDC, and he had acquired his expertise, as he often said, by accident. After attending Western Maryland College (now McDaniel College) on a Reserve Officers' Training Corps (ROTC) scholarship, McDade received a doctorate degree in microbiology from the University of Delaware. He then fulfilled his ROTC service requirement by joining the Special Pathogens Laboratory at the U.S. Army Medical Research Institute of Infectious Diseases (USAMRIID) in Fort Detrick, Maryland. While he waited to complete the immunizations required to work with highly dangerous viruses, however, the laboratory chief switched him to the rickettsial division to fill an unexpected vacancy.

He met his future mentor, Charles Shepard, in 1973 at a conference at the Walter Reed Army Medical Center, where he made a presentation on antibiotic treatment of louse-borne relapsing fever and epidemic typhus. Shepard immediately offered him a job at the CDC and invited him to Atlanta for a visit. By that time, McDade had spent more than a year abroad, first in Cairo and then in Addis Ababa, on contract with the University of Maryland for the U.S. Office of Naval Research. While he enjoyed working abroad—and his studies had been very productive—he was hoping to find a new position in the States where he could settle his family down for the long term. He remembers feeling happy and excited, and full of anticipation, as he walked from CDC headquarters to the hotel across the street after accepting Shepard's offer. But apparently it was not to be. Stopping to pick up a newspaper in the hotel lobby, he noticed a headline that said *"Nixon Freezes Government Hiring."* So McDade returned to Ethiopia and continued his studies on the epidemiology of typhus. One year later, when the hiring freeze ended, he received a new offer from Shepard in the mail.

Ruling Out Q Fever

Fast forward to 1976. Shepard and McDade agreed that Q fever was not a likely cause of Legionnaires' disease because it is associated with domestic animals, such as sheep, cows, and goats. Moreover, it rarely causes severe or fatal illness. But just in case, as part of the intensive search for the cause of LD, McDade went ahead with laboratory testing for *Coxiella burnetti*, the rickettsial species that causes Q fever, employing labor-intensive in vivo techniques that were the standard for rickettsial detection in 1976.

Rickettsia are "obligate intracellular bacteria," which means they cannot live outside of a host cell. When they infect animals or humans, Q fever rickettsiae live inside immune system cells—white blood cells called phagocytic monocytes—that ingest and destroy bacteria, fungi, or parasites. These rickettsiae grow and replicate inside these cells, where they are protected rather than destroyed. Phagocytic monocytes found in the lungs are called alveolar macrophages, because they reside within the alveoli, the tiny air sacs where oxygen is exchanged for carbon dioxide.

Because rickettisa cannot live outside a host cell, they do not grow in liquid bacterial media or on media-impregnated agar in petri dishes. Today, scientists can detect tiny amounts of genetic material, using DNA amplification techniques, bypassing the need to isolate and grow a microbe before identifying it. But in 1976, it was still necessary to grow rickettsia in a susceptible animal (usually a mouse or guinea pig) to obtain sufficient material for testing.

McDade's drill for detecting *Coxiella burnetti* went like this:

- Inject lung tissue from an autopsy of a person who died of LD into the peritoneal cavity of a guinea pig. (The peritoneal cavity is the space within the abdomen that contains the intestines, stomach, and liver.)
- Wait two weeks for the ricketssia to grow, while monitoring the animal's temperature. (A two week incubation period is long for most bacteria, but not for Q fever rickettsiae, which grow very slowly.)
- If the animal runs a fever, sacrifice it and make a suspension from its spleen, an organ that contains white blood cells (including phagocytic monocytes)
- Subculture the spleen tissue in embryonated hen's eggs, adding antibiotics that do not affect rickettsiae (such as gentamicin, erythromycin, or amoxicillin) but will kill off any bacterial contaminants that might be present.
- Monitor the embryos daily until they die from the infection. Then smear material from the egg yolk sac on slides and stain them with Gimenez stain, a special dye for rickettsiae. If rickettsiae are present, they will be visible under the light microscope, living inside the yolk sac cells.

McDade inoculated the guinea pigs and waited for fever to develop, expecting a two-week wait if Q fever rickettsiae were present. But fever appeared much more quickly, in two to three days. McDade sacrificed the guinea pigs, passaged spleen suspensions in embryonated eggs (with antibiotics), and waited again, to see if the embryos would die—but none did. He also smeared the spleen suspension on slides and rinsed them with Gimenez stain: No rickettsia. While examining the slides treated with Gimenez stain, McDade occasionally thought he saw a rod or two—a common bacterial shape—but assumed these were contaminants: single, stray organisms rather than a growing cluster of pathogenic organisms. Nevertheless, just to be sure, he performed a few more procedures to check for other (non-rickettsial) types of bacteria. First, he cultured the spleen suspension in bacterial medium, in addition to embryonated eggs, but nothing grew. Then he made additional slides and treated them with Gram stains (the gold standard for detecting bacteria) instead of Gimenez stain—still nothing. So: no bacteria and definitely no Q fever rickettsia.

No surprise there, because Q fever was a long shot. Negative results were coming in from other CDC laboratories, and the idea of a toxic chemical as the cause of LD looked more and more attractive. In late August, as McDade was finishing up his Q fever experiments, he heard about the nickel carbonyl hypothesis. Since nickel carbonyl is odorless and breaks down quickly, it wasn't so surprising that it hadn't immediately been detected by the CDC toxicology laboratory. McDade thought "yes, that must be the answer"—and folded up. He returned to his interrupted experiments on the rickettsia that cause typhus.

No Solution in Sight

Throughout the fall of 1976, the unsolved mystery of Legionnaires' disease remained in the news and on people's minds. Doctors' offices in Pennsylvania and beyond were flooded with calls from anxious patients, even though the LD outbreak was, in fact, already over. The CDC, which usually maintained a low profile, out of the DC spotlight, was in the news nearly every week, under fire for its decision to implement the swine flu vaccination campaign (which began in October 1), but also because of its failure to find a cause for LD. After all, the CDC had pulled out all the stops—and found nothing. The CDC had fielded the biggest investigative team ever, with more than thirty federal infectious disease and environmental health specialists—including old hands as well as young EIS officers like Marks—assisted by hundreds of state and city public health workers and more than forty city police detectives. And what was the conclusion? A non-contagious airborne substance, microbial or chemical, associated with the lobby and front sidewalk of the Bellevue-Stratford Hotel—a vague and unsatisfying dead end.

Criticism of the CDC was unrelenting, from the press, from Congress, and even from some state and local health departments. On August 31, the Director of the Pennsylvania Department of

Health—the CDC's major partner in the LD investigation—lambasted the CDC for its failure to provide leadership to the swine flu vaccination program, which fell short of its goals because many people refused to get vaccinated. Senator Richard Schweiker of Pennsylvania asked the CDC for weekly updates on the LD investigation and publicized each negative report. His Congressional colleagues, in the heat of a presidential election year, held a succession of hearings at which the CDC was denounced as incompetent, and the swine flu campaign was derided as a waste of taxpayers' money. Nevertheless, President Ford continued to champion the swine flu vaccination program.

In September, the CDC director and the Pennsylvania State health commissioner asked a group of experts to review the pathology data and look for additional clues. The experts included medical examiners from Philadelphia and Allegheny Counties, a professor from the Pathology Department of the University of Pennsylvania medical school, three staff members from the Armed Forces Institute of Pathology, three experts in forensic medicine, and three physicians with special expertise in the pathology of the lung, liver, and kidneys. They reported that chest x-rays of the earliest LD cases revealed patches of scarring and inflammation of the alveoli, the tiny air sacs of the lungs—a condition called interstitial pneumonia that usually indicates a viral infection. On the other hand, the autopsy reports on some of the LD patients showed fatty deposits in the liver, which might indicate a poison or a chemical toxin. (But liver damage might also simply signify excess alcohol intake by the celebrating conventioneers.) With all common causes of pneumonia already ruled out, the experts concluded that the CDC should look for a new virus or an unusual chemical.

There was one, lonely dissenting voice at the CDC, a virologist named Frederick Murphy, an expert on viral hemorrhagic fever who later became head of the CDC National Center for Infectious Diseases. Murphy thought that LD might be a bacterial disease because he saw bacteria when he examined autopsy lung tissues under the electron microscope. However, the chief CDC pathologist was convinced that these bacteria represented a clinically insignificant contaminant.

In October, the swine flu vaccination program came under fire when three elderly people died on the same day, within hours of receiving swine flu shots from a clinic in Pittsburgh. Although the deaths might have been related to improper injection or infection control practices at the clinic, or due to chance, the deaths touched off a temporary suspension of swine flu vaccination in ten states. The vaccination program continued to be unpopular with many groups, although participation among the elderly continued to be high, perhaps because older people remembered friends and relatives who had died during the 1918–19 pandemic.

In November, Congressman John M. Murphy of New York, the chairman of the Congressional Committee for Consumer Protection, claimed that the CDC had bungled the LD investigation by assuming that the cause was an infectious microbe and missing important clues that might have led to the detection of a chemical toxin, like nickel carbonyl. He apparently didn't realize that the nickel carbonyl theory had already fallen apart. On repeating his analyses with additional (non-LD) controls, the University of Connecticut pathologist had realized that nickel was present in all of the samples (controls and non-controls), apparently leached from the scalpels used to perform the autopsies.

By mid-November, the LD investigation, grinding on ever more slowly, nearly came to a halt. Although CDC maintained a hotline that welcomed suggestions from anyone—lay person, politician, or scientific expert—no new theory emerged to intrigue the scientists or energize the rumor mills. The seventy-two-year-old Bellevue-Stratford Hotel closed on November 18, a casualty of the outbreak, a few days after hosting an American Lung Association symposium on the possible causes of LD.[5]

Worst of all, during the third week in November, doctors who vaccinated patients against swine flu began reporting an unusual side effect called Guillain-Barré syndrome (GBS)—an autoimmune condition sometimes triggered by infection. GBS causes muscle weakness that begins in the legs and gradually extends to involve the arms and upper body. Although the weakness usually passes within days, in severe cases it can cause paralysis and death. The vaccination campaign—which had reached about forty-five million people (24 percent of the population) in

10 weeks (a record for a U.S. vaccination program, though far short of what was planned)—was cancelled on December 16.

The Hook

Meanwhile, back at the laboratory, absorbed in his studies on epidemic typhus, McDade was oblivious to the ongoing turmoil over LD and swine flu. He had lost the habit of TV-watching while living in Ethiopia and didn't follow the TV news or the tabloid press. Moreover, he thought of himself as a bottom-of-the-rung, worker-bee scientist who left office politics and the handling of higher-ups to Shepard, his boss and mentor. After all, he was only very peripherally involved in the LD investigation, with just a minor supporting role. Very likely, however, the main thing that insulated McDade from the political heat and media frenzy was his own inter-directed, doggedly-focused scientific mind, which screened out all distractions and focused on the scientific problem at hand (in this case, methods for the detection of typhus).

Nevertheless, from time to time, when McDade came up for air after completing a round of experiments, he had a little, niggling thought about the tiny rod-shaped bacteria that showed up on his Q fever slides—the "hook," as he thought of it, on which his thoughts were snagged. Those rods—could they have meant something after all?

Fast forward once again, this time to the end of November, past the presidential election in which Jimmy Carter defeated Gerald Ford, and past the troubling discovery that the swine flu vaccine was apparently triggering GBS in some vaccine recipients.[6] McDade read a CDC report on LD prepared by David Fraser and an EIS officer named Theodore Tsai that laid out all the theories and all the investigative data. Fraser and Tsai were unhappy about the lack of a solution and the constant criticism, and they were worried about CDC morale in the field and in the laboratory. They submitted the report to the Pennsylvania Department of Health, to the American Public Health Association, and to the CDC director. They wanted to think things over from the beginning, one more time.

By December, McDade got to thinking more and more about those damn rods . . . maybe, maybe, maybe. It was really starting to bother him. He re-read the dispiriting follow-up report. He pretended to ignore a drunken neighbor at a Christmas party, who ragged him about the CDC's dismal failure. But for some reason, these things really got to him. He was angry and shaken because he knew how hard everyone had worked for months and months. There had to be a solution. He thought some more about those rods.

McDade felt more and more compelled to do something, *anything!* Looking back he insists it was not inspiration that drove him, but compulsion. He had to go back and look at those rods once again. He decided to make himself stop what he was doing (a whole other set of typhus experiments) and re-focus on LD. He knew there was little chance that he would find anything that his colleagues had missed, but he was more and more bothered by the problem, almost to the point of obsession. Instead of worrying himself to death, he decided, he would "clarify the issue" one more time and then forget about it.

The Solution

It was the quiet week between Christmas and New Year's. McDade's custom was to use that week of work to "clean house" for the year, while no one else was around. He would finish up his paperwork and tidy up loose ends, undisturbed by colleagues or obligations.

On his own in the laboratory, he took out the box of guinea pig slides from August, the ones he had stained for rickettsiae by the Gimenez method. He viewed each one, section by section, very slowly, under the light microscope. Here and there he found a rod-like shape, just as he remembered. But then—there it was—he found a slide that had not just one rod, but a whole cluster of rod-shaped microorganisms. He went to the freezer, where he had carefully stored material from the first set of guinea pig experiments. He located a spleen-cell suspension from one of the guinea pigs that had been inoculated with lung tissue from a deceased Legionnaire. Then he sat down at his desk and thought over what he should do to re-evaluate his previous findings when he returned to the lab after the Christmas holiday.

He began on December 26. As part of his Q fever experiments, he had inoculated the guinea-pig-spleen suspension into embryonated eggs to see if anything would grow out, but nothing had. This time, he did the same procedure, but left out the bacteria-killing antibiotics. This time the embryos died in a few days. McDade smeared yolk sac tissue onto slides and stained them with rickettsial (Gimenez) stain. This time, they were loaded with rod-shaped bacteria! Was this evidence of contamination, or did it mean something?

He thought of an easy test. (*Easy* in the sense of simple and conclusive, not *easy* in the sense of not a lot of work.) He would find out whether LD patients had antibodies for the yolk-sac bacteria in their blood, using a standard laboratory technique called the indirect immunofluorescent assay (IFA). (Antibodies are special proteins in the blood produced by the immune system in response to microbes and other foreign bodies. Antibodies adhere to the microbes and foreign bodies, flagging them for ingestion by phagocytic cells.) He went to find Gary Noble, the head of the CDC influenza laboratory, and asked him for blood samples from the LD case-control study, including both case-patients and controls. If the bacteria in the yolks were the cause of LD, antibodies in blood samples from the LD patients—but not the control samples—would stick to them. It was a stroke of luck that Noble was also in the lab that week; he not only provided the samples but also encoded them so that McDade could test them without knowing which were from LD patients and which were from controls.

With the assistance of his long-time technician, Martha Redus, who returned to the laboratory after New Year's Day, McDade smeared multiple spots of yolk sac tissue from the "no-antibiotics" experiment on microscope slides and used acetone to fix the tissue to the glass. Next, he placed the slides on a tray and added a drop of serum (the part of blood that contains antibodies) to each slide. Each drop of serum was from a different coded (LD or non-LD) blood sample. He placed the tray in an incubator for an hour to give the antibodies time to find and stick to the bacteria. Then he rinsed off the slides. The next step was to add a fluorescent reagent that would adhere to the antibodies stuck to the bacteria and be visible under a UV microscope. For this purpose he used "goat anti-human antibody"—a goat antibody that adheres to human proteins—attached to a fluorescent dye.

It was now early evening. McDade cranked up the UV microscope. Some of the slides lit up like crazy—apple-green bright! The hair on his neck began to prickle. McDade asked Redus to record the results. Then, they decoded the tests using Noble's crib sheet. The positive slides were the ones from LD patients! All the other specimens were stone-cold negative. Wow! But what did it mean? They repeated the whole thing, starting this time with a spleen suspension from a guinea pig injected with lung tissue from a second LD patient.

This experiment—starting again with egg culture—took a few days. Once again, all of the LD slides lit up, while all of the controls remained unstained. But he was still afraid to conclude that he had succeeded where everyone else had failed. So he tried a more definitive (and quantitative) test that made use of paired blood samples from LD patients who had participated in the case-control study. The first sample in each pair was taken during the initial onset of disease (when an immune response had barely begun) and the second during the convalescent period (when the blood serum would be full of antibodies), two or three weeks later. McDade hoped to see an increase in the concentration of antibodies (the antibody titer) when he compared each pair of onset and convalescent serum samples. A rise in antibody titer in recovered patients (but not in healthy controls) would be fairly convincing evidence that the bacteria that induced the antibodies had caused the disease.

McDade used Teflon plates with ten wells each. He put yolk sac material in the first well, diluted it by half in the second well, half again in the third well, and so on. He then repeated the earlier series of steps, adding a drop of serum to each well, incubating the plates, rinsing, adding the fluorescent probe, incubating, and rinsing again. For this experiment, two drops of serum were tested from each patient, one taken from the disease-onset serum and one from the convalescent serum. He focused carefully on each step, trying not to anticipate what he would feel when he saw the results—relief and happiness, or maybe a tremendous let-down (a common feeling to those who work in the laboratory). By late evening, the results were in. McDade asked his technician to check the first pair of samples. The first one (the onset sample) had little or no titer reaction, but the second one (the convalescent sample) had a high titer, and the titer was lower by half in each successive well (512, 256, 32, etc). Once again, the hair on his neck began to prickle. . . .

Of thirty-three paired samples from recovered LD patients, twenty-nine showed a positive reaction ("seropositivity") in one or both samples. In nineteen cases, there was at least a four-fold antibody rise between the onset sample and the convalescent sample, indicating that the immune reaction rose rapidly in the second and third weeks of illness ("seroconversion").

By this time it was late at night, and McDade and his technician—excited and amazed—went home to sleep on it. It seemed that they had solved the puzzle, but if they claimed success without being one hundred percent sure, they—and the CDC—would be in big public trouble.

Early the next day, McDade shared his observations with Shepard, who responded with his typical calm and cautious demeanor. A very interesting finding, but given the relatively small number of specimens tested, they could not be sure they had identified a new organism. They discussed alternative explanations, such as contamination (an ever-present concern) and the possibility that the new microbe was an innocuous, non-pathogenic fellow-traveler of the actual causative agent of Legionnaires' disease. After much discussion, they decided to repeat the entire experiment, beginning with lung tissue from a patient who had died from Broad Street pneumonia rather than LD. Would they find the new microbe in Broad Street patients as well? The answer, which took another week of injecting, waiting, subculturing, waiting, smearing, and staining was a resounding *yes*.

It was now Friday, January 14, 1977. Shepard decided it was time to speak to the CDC director, David Sencer, who was badly in need of some good news. He ushered McDade into Sencer's office, where McDade tried to remain in the background while his boss did the talking. He felt like a rookie in the presence of these two distinguished scientists, who were accompanied by other senior CDC staff members. Sencer reviewed the evidence that the causative agent of Legionnaires' disease was a bacteria—an unusual bacteria that didn't grow on standard media or in mice. He asked questions, and Shepard gave most of the answers. The data looked good, so far. What did they plan to do next?

Shepard wanted the experiments to be repeated in another facility, so there could be no question about the introduction of an extraneous contaminant. The next step would be publication in a scientific journal that would send the article for external "peer-review"—professional critiques by microbiologists who might think of alternative explanations and suggest additional tests. But the besieged CDC director had a different idea. He suggested moving things along more quickly by publishing the results in a special edition of the CDC *Morbidity and Mortality Weekly Report (MMWR)*, the foremost record of public health news, read religiously by epidemiologists. It was essential, he said, to inform health officials and the public of these new developments in a timely manner. Shepard and McDade could have the weekend to repeat any experiments they deemed necessary (along with any new controls they might think of) as long as they were ready to publish on the following Tuesday (normal *MMWR* publication was on Fridays). The *MMWR* article would include the caveat that the newly identified bacterium, while associated with LD, might be a secondary invader. It would acknowledge that additional studies were required to characterize and classify the microbe, which appeared by size, shape, and antibiotic sensitivity to be a non-rickettsial bacterium.

Over the weekend, Shepard had an inspiration. Early Monday morning he asked his lab technicians to locate paired serum specimens from two earlier unsolved outbreaks (in 1966 and 1968) that had always troubled him. He remembered that the victims of the 1966 outbreak—residents of St. Elizabeth's Hospital for the Mentally Ill in Washington, DC—had LD-like respiratory symptoms, including pneumonia. More than ninety people had fallen ill, and sixteen had died. Two summers later, there was another puzzling incident that affected nearly all the employees of the county health department in Pontiac, Michigan, who were sickened by a non-contagious illness whose symptoms included fever, headache, and muscle pain. Although the symptoms were not LD-like (none of them had pneumonia and all of them recovered), the outbreak was associated with a single building (a downtown office building) that had a central air-conditioning system of the same type as the Bellevue-Stratford. Shepard recalled that several members of the CDC investigative team in Pontiac had contracted the mystery disease after they had turned on the building's air conditioner, which had been turned off when they first arrived.

When McDade tested serum samples from the St. Elizabeth's outbreak on the yolk-sac bacteria, the slides lit up once again! Was this in fact a common but undiagnosed disease? On Tuesday morning, Sencer agreed to stop the *MMWR* presses and add the St. Elizabeth's data to the special edition of *MMWR*. (The Pontiac serum specimens subsequently tested positive for LD antibodies too.)

On Tuesday afternoon, before holding a formal press conference, Sencer got on the phone with a large group of public health professionals, including state health officers, the Surgeon General, and the Director of the National Institutes of Health, to gave them a heads-up. Shepard and McDade requested that all CDC employees who had worked on the LD investigation, from glassware handlers in the laboratory to the chiefs of epidemiology, be invited to participate in the call.

More Pieces of the Puzzle

A new epidemiologic team was immediately dispatched to Philadelphia to look for evidence of a connection between the new organism and the outbreak at the Bellevue-Stratford. The team's first objective was to take new water samples (filtered and unfiltered) from the cooling towers on the roof and from the pipes that fed into the taps in the main reception room. Lo and behold: the non-filtered samples from the cooling towers were positive for the new organism, using McDade's guinea pig assay. Though they could not prove that the bacteria came from the air conditioner—because the filters had been cleaned or changed before the investigators arrived—the circumstantial evidence was strong: the bacteria were living in the cooling towers that fed the air conditioner above the registration desk in the lobby—the same cooling towers that sprayed aerosolized water onto the sidewalk in front of the hotel. It seemed plausible that the worn-out filters had been cleaned or changed too late to prevent the bacteria from entering the water in the cooling towers. But soon after the filters were taken care of, the outbreak had stopped.

Back at the laboratory, the reinvigorated CDC researchers quickly figured out why the bacteria had been so difficult to detect. It turned out that the LD-associated bacterium does not grow on standard media because it has special dietary requirements, including extra-large amounts of iron and the amino acid cysteine. Moreover, it does not absorb common bacterial stains, either because of the properties of its cell wall, or because (like rickettsia) it lives sequestered inside alveolar macrophages. Apparently for the same reasons, it does not take up most antibiotics that might otherwise kill it or prevent its replication.

After considerable testing, the scientists found an antibiotic—erythromycin—that could diffuse into phagocytic cells and kill LD bacteria. They also devised a special stain for LD bacteria, even more effective than Gimenez: a modified form of a fifty-year-old silver-impregnation technique used to view spirochetes, the causative agent of syphilis. Spirochetes—a type of bacterium with a distinctive helical morphology—also spend part of their lifecycle inside human cells. Finally, they developed a culture medium that contained double the amount of cysteine and iron found in standard bacterial media.

Armed with the new culture technique and silver stain, the CDC scientists demonstrated unequivocally that the new microbe was a gram-negative bacterium rather than a ricketssial bacterium, although it turned out to be closely related genetically to *Coxiella burnetii*—the rickettsia that causes Q fever. (Gram-negative bacteria are characterized by an inability to take up certain dyes because of the structures of their cell walls.) It was named *Legionella pneumophila* ("lung-loving"), in honor of the American Legionnaires. Although many affected groups do not want the stigma of having an organism or a disease named after them, the leaders of the American Legion decided that the name would honor their fallen colleagues.

There was soon more evidence that *Legionella* was not, after all, a new pathogen, but an old one that had never been detected before. Working through the archived specimens in the CDC deep freeze, McDade and his colleagues found evidence of additional LD cases from as long ago as 1943. Marilyn Bozeman, a biologist at the FDA, sent McDade four rickettsia-like bacteria she had isolated in 1947, using the same techniques McDade used to isolate *Legionella*. Although none of them grew in the new LD medium, McDade went ahead anyway and tested all four by the guinea pig assay. One of them—now called *Legionella bozemanni* in Bozeman's honor—proved to be another pathogenic species of *Legionella* with different antigenic properties and dietary requirements. Moreover, one of the deep-freeze specimens (from 1954) was an unusual *Legionella* species that lives in free-living protozoa in the soil in Australia, where it is a danger to gardeners. All other species in the family Legionellaceae—and there are now more than fifty—are found in water.

It turns out that the natural habitat of *Legionella* is not humans or animals, but rather rivers, creeks, ponds, and other freshwater environments all over the world. The bacteria flourish in riverbed mud, as well as in "pond scum"—the greenish film on the surface of freshwater ponds. In ecologic terms, pond scum is actually a "biofilm," a floating layer of free-living green algae that supports many other organisms. *Legionella* spend at least part of their life cycle inside single-celled organisms—protozoa—that live on the algae in the pond scum, which are a good source of iron and cysteine. The *Legionalla* reproduce inside the protozoa, and one protozoa, carried in an aerosolized drop of water, contains enough bacteria to infect a human being.

For unknown reasons, human infection with *Legionella pneumophila*—the *Legionella* species responsible for the great majority of legionellosis cases in the United States—can give rise to two different clinical manifestations: A mild, self-limited, flu-like illness, called "Pontiac fever," and Legionnaires' disease, which can cause severe illness, involving pneumonia. Experts estimate that eight thousand to eighteen thousand Americans are hospitalized because of LD each year, although few of them are diagnosed as LD. The LD fatality rate is about 15 percent, with those at most risk being elderly people, smokers, people with chronic lung disease, and people immunocompromised by chemotherapy or HIV infection. Antibiotic treatment with erythromycin is usually effective, especially if given early.

McDade calls LD a "disease of modern technology," because *Legionella* in their natural settings—waterfall, lakes, streams—do not cause illness. In fact, *Legionella* only become sufficiently concentrated to cause disease when present in stagnant water that is very hot (77–108 °F) and aerosolized, which is rarely if ever the case in rivers and streams. Human disease is almost always associated with human-made devices that contain warm water—including hot tubs, water heaters, whirlpool spas, humidifers, Jacuzzis, drinking fountains, and cooling towers—which provide ideal surfaces for the growth of biofilms. In fact, the introduction of air-conditioning systems cooled by circulating water during the 1940s may have made human infection with *Legionella* much more common. Moreover, the practice of coating air conditioning filters with motor oil—a standard maintenance procedure at the Bellevue-Stratford—apparently created a nutrient-rich, pond-scum-like biofilm that was especially conducive to the growth of *Legionella*. In addition to outbreaks associated with aerosol-producing devises (most common in spring and summer), some outbreaks have also been associated with construction sites, possibly due to disruption of biofilms inside pipes that causes *Legionella* to leach from the biofilms into the water flowing through the pipes.

A drop of contaminated pond water that makes its way to a hot tub or cooling tower (e.g., through the air or on a person or animal) may be sufficient to seed a new *Legionella* habitat. In turn, an inhaled drop of contaminated water from the tub or cooling tower can enter a person's alveoli, take refuge in an alveolar macrophage—which serves as a substitute for its usual protozoan host—and begin to multiply and destroy alveolar cells

The detection of *Legionella* bacteria in cooling-tower water from the Bellevue-Stratford and later from many other devices—including an ultrasonic machine used to spray mist over a display of vegetables in a grocery store—led manufacturers to issue new guidelines for routine maintenance procedures for cooling towers and other aerosol-producing devices. It also led to use of monochloramine instead of chlorine in some municipal water systems and hospital water systems because monochloramine is better able than chlorine to penetrate biofilms and kill *Legionella*.

Lessons Learned

McDade's discovery of the causative agent of LD in January 1977 occurred a month after the cancellation of the swine flu vaccination program and a month before David Sencer—a highly respected public health expert whose eleven-year tenure as CDC director spanned the Johnson, Nixon, and Ford administrations—was fired. He had tried to avert a disaster that had not materialized. Afterwards, the CDC experienced a period of intensive soul-searching about the handling of the swine flu threat.[7] In contrast, the lessons learned from the LD investigation seem positive and straightforward. The LD investigation is a good illustration of how the two interdependent cadres of the public health system—epidemiologists and microbiologists—work in parallel to solve an infectious disease mystery. Though McDade's expertise was in bacteriology, he used epidemiologic data to crack the case. Today, the laboratory investigation would no doubt move faster, due to the availability of DNA amplification techniques based on the polymerase chain reaction (PCR), which can detect tiny amounts of a microbe-specific DNA sequence, even if the microbe is difficult to isolate or grow.

The LD investigation also underscores the importance of the individual in scientific discovery. As in any professional field, some people are motivated primarily by a wish for recognition and outward success, while others are enthralled by a particular subject or the need to answer a particular question. McDade is a perfect example of the inner-directed type of scientist who has a relentless inner need ("a compulsion," as he told me) to understand what's going on. To this day, McDade denies that his pursuit of the mysterious rods had anything to do with scientific inspiration. He remembers telling Larry Altman of The *New York Times* that nobody asked him to do it. Not his boss, not anyone. It was just something he had to do.

In terms of public health policy, a major take-home lesson of the LD investigation is that we should always be ready for the unexpected, since infectious diseases are constantly emerging and "old" pathogens can find new niches, taking advantage of human behaviors and new technologies. In practical terms, this implies that the CDC and its infectious disease detectives need to maintain scientific expertise that is both broad and deep, enabling investigators to rule out known diseases and identify new ones.

The idea that the CDC should maintain a breadth and depth of scientific expertise in all types of infectious diseases was not a popular idea in the 1970s, when McDade solved the LD mystery. In fact, during an award ceremony at the Infectious Disease Society of America (IDSA) in 1977, McDade was told by an officer of the IDSA board that his discovery of *Legionella* was merely an anachronism, because all significant microbes had already been discovered! This remark reflected the common belief that infectious diseases were a thing of the past, conquered by vaccines and antibiotics, and that it was time to shift health resources to chronic diseases and the war on cancer initiated by President Nixon in 1971. As a consequence, the CDC's programs on "low priority" infectious microbes (i.e., those rarely seen in the U.S.) received little funding, and some were nearly dismantled.

The realization that microbes are in fact evolving, adapting, and spreading faster than we can conquer them came in the 1990s, following the worldwide pandemic of HIV/AIDs and the emergence of multidrug-resistant tuberculosis. With support from Congress and encouragement from the Institute of Medicine, the CDC began to rebuild the component of the public health system that protects the public against infectious diseases, by strengthening disease surveillance and diagnostic resources at the state, local, and federal levels. In the mid-1990s, CDC scientists participated in several international outbreak investigations of highly dangerous re-emerging diseases, including Ebola hemorrhagic fever in Zaire (now the Democratic Republic of the Congo), plague in India, and a severe manifestation of leptosporosis in Nicaragua. (Both plague and leptospirosis are animal-borne diseases that had been on the CDC's "low-priority" list.)

The emphasis on broad-based preparedness for the unexpected lasted through the 1990s but was largely abandoned after 9/11 when the CDC's efforts were redirected to what appeared to be potentially imminent threats to "global health security" (e.g., bioterrorism and pandemic influenza). This approach continued through the 2000s, despite the continued appearance of unforeseen infectious disease events (e.g., the emergence and international spread of severe acute respiratory syndrome [SARS] and the appearance in Wisconsin of monkeypox—a relative of smallpox endemic to Central Africa—transmitted by infected prairie dogs sold as exotic pets).

To this day, McDade thinks it was fortunate that Shepard had the foresight back in the 1970s to hire a rickettsial expert who knew how to test for an intracellular pathogen that does not grow in regular medium or in mice. If McDade (or someone with his training) had not been there, the discovery of *Legionella pneumophila* might have been delayed for a very long time.

One more take-home lesson from the LD investigation is that the practice of assigning young epidemiologists to major investigations is an effective way to train and inspire them. Marks agrees that working on the LD investigation was an amazing, life-changing experience. Working on a major investigation, he says, is like working in "the intensive care unit or the emergency room of public health." He remembers the excitement, the feeling of uncertainty and risk, and the confirmation that he had chosen an important, exciting, and satisfying field. Evidently the lack of an immediate solution did not dampen his enthusiasm, or that of other EIS officers involved in the LD investigation, many of whom went on to impressive careers. In addition to Theodore Tsai (now at Wyeth Lederle Vaccines) and Marks himself (senior vice president and director of the Health Group, Robert Wood Johnson Foundation, and former Director of the CDC's National Center for Chronic Disease and Health Promotion), they included David Heymann, Board Chairman of the U.K. Health Protection Agency and former Assistant Director-General for Health Security and the Environment at the WHO; Walter Orenstein, Deputy Director for Vaccine-Preventable Diseases at the Gates Foundation and former director of the National Immunization Program; and Stephen Thacker, who became head of EIS in 1989 and served in a succession of key CDC positions, including Director of the Office of Surveillance, Epidemiology, and Laboratory Services, until his death in 2013.

Marks recalls that on his return to Ohio he was immediately assigned to local investigations, with responsibility for all aspects of the process, from start to finish—gathering epidemiologic data, analyzing it, coordinating with laboratory scientists, and (if the event was ongoing) providing advice on how to prevent further disease spread. To his surprise, before a year had passed, he had helped investigate three cases of respiratory disease at a community hospital in Columbus that turned out to be caused by *Legionella* bacteria![8] Over the following years, as a laboratory test became widely available, additional LD cases were identified across the nation and overseas, confirming that LD is an old disease, newly discovered.

Post-Script

As it happens, *Legionella pneumophila* was not the last pathogen that Joe McDade helped identify. Twelve years after the LD outbreak in Philadelphia, he discovered that another group of rickettsia-like organisms causes a tick borne disease prevalent in the southeastern and south-central United States. This time, his discovery was not made because of an outbreak, but because of a picture. In 1989, a medical pathologist in Detroit contacted a colleague of McDade's for help. He wanted to establish a cause of death for a man who had the symptoms of Rocky Mountain spotted fever but had tested negative for it. The pathologist determined that the patient had recently visited rural Arkansas, where he was exposed to animals that carried ticks. Unable to interest McDade's colleague, the pathologist called the next person on his list—McDade—who listened patiently and asked him to send a micrograph (a picture taken under the microscope) of the patient's blood cells.

The micrograph, which showed lots of red blood cells, but only a few white ones—some of which appeared to have a micro-organism inside them—reminded McDade of an unusual set of slides he had seen many years before, in 1969, after he began his rickettsial studies at Fort Detrick. The slides were presented at a conference at the Walter Reed Army Institute of Research by an army major named David Huxsoll, who was studying a febrile disease that was killing military working dogs (mostly German Shepherds and Labrador retrievers) that functioned as trackers, scouts, sentries, and mine-detectors in Vietnam. The disease, tropical canine pancytopenia, was caused by a tickborne rickettsia of a genus called *Ehrlichia* that can multiply inside an animal's white blood cells and destroy them within a matter of months. All these years later, McDade still remembered Huxsoll's slides, because of the dramatic loss of white blood cells and the striking images of the causative agent.

At the time, however, *Ehrlichia* was generally not thought to cause disease in humans, with a few exceptions.[9] McDade asked the Detroit pathologist to send serum specimens from his patient. He also contacted Miodrag Ristic, a rickettsial expert at the University of Illinois, and asked him for preparations of *Ehrlichia*. Employing the same IFA technique he used in the LD investigation, McDade determined that the patient had indeed developed antibodies to *Ehrlichia*. His experiments set in motion a series of follow-up investigations, at the CDC and elsewhere, that culminated a few years later in the isolation of a new species of *Ehrlichia* that causes human disease—the first such detection in the Western Hemisphere. It turns out that human ehrlichiosis occurs worldwide and can be caused by several different species of *Ehrlichia*.[10]

McDade's microbe-hunting days were still not over. Six years later, in 1993, he helped a CDC team solve another outbreak mystery, which is described in Chapter 7.

In retrospect, it seems evident that the 1977 discovery of *Legionella pnuemophila* was not an anachronism, or a fluke, but rather a harbinger of things to come. Over the past three decades, a resurgent coterie of microbe hunters, armed with modern molecular tools, has identified new pathogens at an astonishing rate—more than thirty in as many years—including bacterial species that cause Lyme disease and peptic ulcer disease, and viruses that cause AIDS, hepatitis C, and SARS.[11]

5

Dangerous Desserts

Medical detective Craig W. Hedberg first learned about the rise in food poisoning cases from a microbiologist at the Minnesota state lab who visited his office in late September, 1994. The cases were all due to a single serotype of *Salmonella*—*Salmonella enterica,* serotype Enteriditis—which causes a miserable mix of abdominal cramps, fever, and bloody diarrhea. Although *Salmonella* was on his department's "watch list," the small increase in this common serotype did not raise a red flag or provoke anxiety in Dr. Hedberg, a sensitive man who appears calm and unflappable. As a general rule, he regarded cases of this particular *Salmonella* serotype as common and unremarkable.

Over the following days, however, the number of cases began to rise, in a slow but continuous way. Most of the reports came from the southeastern corner of Minnesota, a mostly rural area below the Twin Cities that includes scenic small towns along the Upper Mississippi Valley, as well as the city of Rochester and college towns like Northfield and Winona. The clustering of cases suggested a common origin—or point-source—such as a tainted meal served at a church picnic or local diner.

Hedberg noticed that several patients were young adults, which was surprising, because doctors are commonly consulted about diarrheal illness only when a patient is very young or very old. One physician called attention to a young man who had participated in a "Defeat of Jesse James Days" celebration in Northfield, the site of an 1876 bank robbery by the James-Younger Gang. Could this be the point-source—a contaminated food or drink sold at the festival? Hedberg and his staff phoned several patients, but none had visited Northfield, attended a common event, or eaten at a particular restaurant.

The medical detectives considered asking a colleague in Rochester to make house calls to look for unusual food products in the patients' kitchen cabinets. But by this time the laboratory had confirmed more than thirty cases of *Salmonella* Enteriditis, which is well above the normal background level. (In comparison, only ninety-six cases had been reported in all of 1993.) The reports were no longer limited to southeastern Minnesota, and the number of cases was not leveling off. Something was going on—something bigger than a church-picnic-type event. Hedberg decided to figure it out.

Foodborne diseases, which cause millions of illnesses and thousands of hospitalizations and deaths in the United States each year,[1] are not easy to track or investigate. They are vastly underreported, because people with diarrhea rarely seek medical care, and when they do, their doctors seldom send stool samples for laboratory testing. As a result, the cause of most outbreaks (e.g., a bacterium like *Salmonella* Enteriditis or a virus or parasite) and its source (e.g., chicken, mangoes, eggs, or lettuce), often remains unknown. Sometimes a diagnosis of salmonellosis is laboratory-confirmed when a patient is hospitalized with severe diarrhea or complications, such as a bloodstream infection that could affect the brain, spinal cord, heart, or bones.

Although Hedberg was starting to worry, he knew exactly what to do, because his department—the Minnesota Department of Health (MDH) in St. Paul—had a decade-long history of aggressively investigating outbreaks caused by contaminated food. His boss, Michael Osterholm, the State Epidemiologist of Minnesota, had made foodborne diseases a special priority. Under his leadership, clinical labs at hospitals and clinics throughout the state—which in 1994 had 4.5 million people—were encouraged to send clinical isolates to a single state laboratory in Minneapolis for confirmatory testing and subtyping of foodborne pathogens such as *Salmonella, Listeria,* and *Escherichia coli.* When the results were sent back to the hospital or the patient's doctor they were also reported to MDH epidemiologists, who compiled the data over weeks and months and used it to search for

The McConnon family, at home in Bangkok in 1980. From left to right: Jessica, Patrick, Kenevan, and Kate.

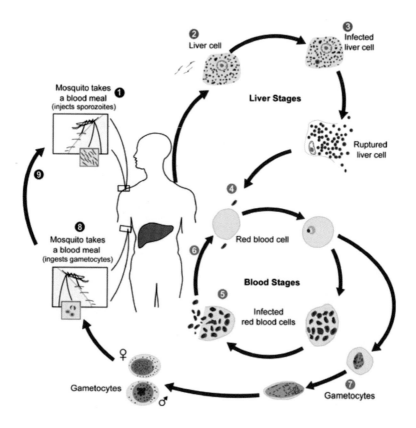

Life cycle of malaria parasites. Malaria is caused by a single-celled organism with a complex life cycle that includes an infectious-stage parasite (the sporozoite; introduced into the bloodstream by mosquito bite); a liver-stage parasite; and a blood-stage parasite, which causes the classic malaria symptoms of fever and chills. Diagram courtesy of CDC-DPDx.

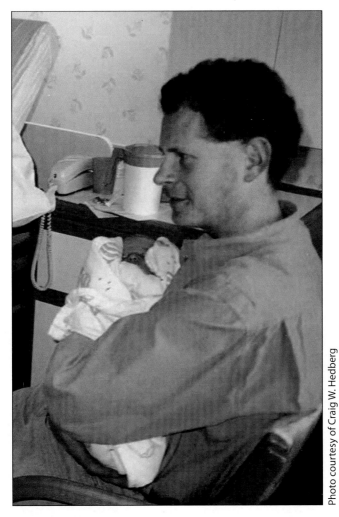

Medical detective Craig W. Hedberg with his newborn child.

Medical detective Thomas Hennessy carrying
the floor plans of the Schwan ice cream factory.

A tanker truck that hauled pre-mix to the Schwan ice cream factory.

Minnesota medical detectives investigating the outbreak at the pork processing plant. From left to right: Stacy Holzbauer, Ruth Lynfield, Richard N. Danila, and Aaron DeVries.

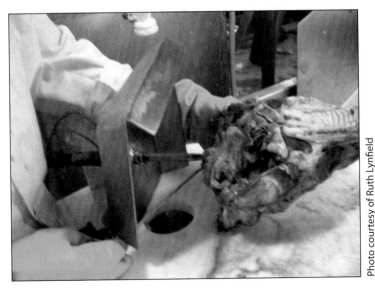

Demonstration of how a swine brain can be removed from its skull using a pressurized air device.

Rodent Processing Before and After the Discovery of Sin Nombre Virus

Photo courtesy of Jamie Childs

Before

Photo courtesy of Kenneth Gage

After

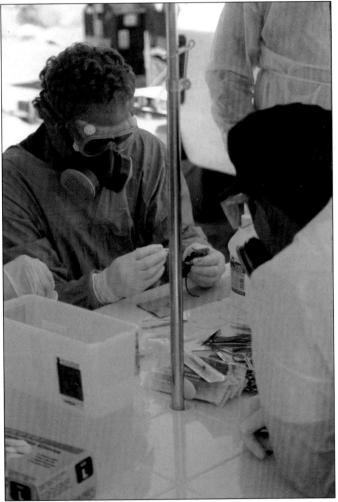

Jamie Childs processing rodents during the field investigation in Four Corners.

A deer mouse captured in Four Corners.

Immunofluorescent staining of human primary lung microvascular endothelial cells infected with Sin Nombre virus (SNV). The cells were stained using anti-SNV rabbit polyclonal antibody (green), anti-actin mouse monoclonal antibody to detect the cell's cytoskeleton (red) and DAPI stain (blue) to detect the nucleus.

upward trends that might represent the tip of the iceberg—indicators of something big. Prompt investigation of these trends—made possible by a sustained commitment of precious public health resources—had allowed MDH to control foodborne outbreaks that would have otherwise remained below the radar.

Osterholm likes to say that "Mother Nature gives us opportunities to learn" and that knowledge can be gained from each investigation. Over the years, he had encouraged his staff to build the public health literature on foodborne diseases by publishing their findings in professional journals. From one investigation they learned that tomatoes are not too acidic to provide a good environment for *Salmonella*, as formerly believed.[2] From another they learned that cheese with a barely detectable level of *Salmonella* contamination can cause a significant outbreak.[3] In 1985, soon after Hedberg joined the department, the MDH detectives, working with colleagues from South Dakota and the CDC, discovered that a cluster of cases of what appeared to be a chronic condition—hand tremors and muscle weakness caused by an overactive thyroid—was actually due to eating a contaminated food: hamburger made with extra-lean ground beef that included cow thyroid glands.[4] Hedberg remembers working long but productive hours to make sure more people did not become sick. He also recalls that once the tainted meat was off the market and things began to wind down, MDH shut down the investigation for a week to give him time off to get married. A few years later, this unusual foodborne outbreak became the topic of the last *New Yorker* article by Berton Roueché, the author of *Eleven Blue Men,* a classic collection of outbreak stories from the 1940s and 1950s.[5]

So now Hedberg was ready to "pull out all the stops," as his staff called it, mobilizing the entire department and setting in motion well-exercised procedures for tracking down a source of contamination. The first step was to compare the foods eaten by people who fell ill ("case-patients") and those who remained healthy ("controls"). The case-patients' medical histories had to meet a stringent case definition to ensure that only outbreak-related cases were included, and the controls had to be matched to the patients by age, sex, and neighborhood (as determined by the person's telephone number). Hedberg's team made call after call until a good match was made for each patient. Many of them were teenagers or young adults, as Hedberg had observed.

They began the case-control study on Wednesday, October 5. By Thursday afternoon they had interviewed fifteen patients and fifteen matched controls, using an exhaustive foodborne disease question-naire. They asked which foods each person had consumed, at home or at a restaurant or party, within five days of falling ill. (If a patient was very young, they interviewed the child's parents.) The idea was to stir memories and prompt people to recall whether they had eaten specific food items, including meat, eggs, dairy products, store-bought fruits and vegetables, and processed foods. It was a long list, because *Salmonella* can grow on just about any type of food, although eggs are the most frequently identified source for *Salmonella* Enteriditis. What did you have for breakfast, lunch, and dinner each day? What kind of meat? What brand of cheese? What flavor of yogurt? Where was it bought? When was it eaten? The team asked each person to consult checkbooks, menu plans, and grocery receipts.

Hedberg made some of the calls himself. Two of the first three patients mentioned a food product he had never heard of—ice cream made by a local company called Schwan's. The newest member of his staff, Thomas Hennessy, filled him in. Hennessy, a brand-new trainee from the CDC Epidemic Intelligence Service, explained that the Schwan Food Company was a popular source of frozen food in many rural areas, including the Four Corners area of New Mexico where he had spent four years as a family physician. (Hennessy's experience working on the Navajo Reservation in Four Corners is described in Chapter 7.) Schwan's products were delivered door-to-door every week by friendly salesmen, like the Fuller Brush Man or the Avon Lady. The bright yellow ice cream trucks, with their dependable routes, were a happy symbol of ice cream and com-fort—a rural counterpart to the Good Humor truck that went from playground to playground in many U.S. cities. As a doctor in Four Corners, Hennessy had carried countless lunches of Schwan's pizza and popsicles to hospitalized children.

Ice cream is a favorite food item that people do not think of as hazardous—a special American comfort food. Could ice cream be the source of the outbreak, or was it another red herring, like the Jesse James event? In little more than twenty-four hours, Hedberg had an answer: Eleven of the fifteen patients (73 percent) but only two of the controls (16 percent) had eaten ice cream products made by the Schwan Food Company. The results were clear and convincing. Although fifteen is a small number, the fifty-seven point spread between 73 percent and 16 percent is significant by any statistical test—too large to be due simply to chance. Moreover, the eleven patients who had eaten Schwan's ice cream had enjoyed a variety of products, including ice cream of different flavors, dietary desserts, ice cream cones, frozen yogurt, and ice cream sandwiches. That suggested a much bigger problem than a single bad batch of ice cream.

Osterholm agreed that the data were compelling and that action had to be taken to prevent additional cases. An ice cream outbreak is not like an outbreak caused by contaminated fruits or vegetables that flares up and is over. Ice cream stays in the freezer over weeks or months without spoiling and is a good medium for preserving bacteria. Osterholm spoke with colleagues at the Minnesota Department of Agriculture (MDA) and the local office of the Food and Drug Administration (FDA). He also informed his boss, the State Minnesota Commissioner of Health, Mary Jo O'Brien, and arranged to meet with Schwan Food Company officials on the following day, Friday, October 7.

Hennessy remembers that the executives from Schwan's arrived at MDH "fully armed" with a squad of lawyers. The company's microbiologist was skeptical. He had tested all of the flavorings and ice cream products, as he always did, and found no problems. While he agreed that his assays might miss a very low concentration of bacteria, they complied fully with industry requirements. The FDA always required direct evidence of *Salmomella* contamination before taking action, and no bacteria had grown out on his petri dishes! The FDA had never recalled a product for *Salmonella* contamination on the basis of an epidemiologic study alone.

Osterholm stood his ground, insisting that the company act without waiting for laboratory results. It was too risky not to do anything, because tainted ice cream was sitting in people's freezers and more kids could get sick. He was going to issue a public announcement that evening, in time for the evening TV news, whether or not Schwan was on board.

Hennessy was impressed. Osterholm was going out on a limb to protect the health of the people of Minnesota! Here was an elegant example of using data to drive public health action—the medical detective's greatest ideal. Hennessy was also happy to hear the MDA official concur with Osterholm. The MDA official had worked with Osterholm on a previous case, involving contaminated cheese.[3] Osterholm had stuck to his guns then too—and he had been right.

Later that day Commissioner O'Brien spoke by phone with a Schwan Food Company official. A staff member, Richard Danila, was with her when she made the call. Danila was in charge of Minnesota's HIV/AIDS unit, a position that required political savvy and PR skills as well as scientific expertise. He remembers standing in the room, wondering how the Commissioner would handle the call. To his surprise and relief, she backed up Osterholm completely. She said that a press release would be issued that afternoon, either by Schwan or by MDH, to warn the public about the ice cream.

A few hours later, the Schwan Company complied, announcing its intention to conduct a voluntary recall of its ice cream products, not only in Minnesota, but all over the country. It closed down its primary production facility, in Marshall, Minnesota, and asked its home-delivery customers (by letter and through advertisements) to discard or return all uneaten ice cream. Schwan's even sent their trucks door-to-door to retrieve ice cream from each household. The trucks retrieved and dumped a huge and impressive amount of ice cream.

Schwan's public awareness campaign—advising its customers not to eat its ice cream—was unprecedented and has been praised and studied as a model of good corporate citizenship.[6] The company protected its customers and appeared to cooperate fully with the public health investigation. However, behind the scenes the situation was more complicated, because not everyone at Schwan initially agreed that a recall was necessary. Despite its October 7 announcement, Schwan did not instruct its drivers to implement the recall until Sunday, October 9.[7] Moreover, in a series of letters to the Minnesota State attorney general's office, MDH, the CDC, and the FDA, Schwan's lawyers disputed health officials' requests for information and tried to limit what Osterholm could say to the press.[7] The lawyers claimed that Schwan's nationwide production and distribution data—requested by MDH and the CDC to assess the scope of the outbreak and calculate the number of people at risk—was proprietary. Moreover, Schwan did not want Osterholm to reveal his estimate of the number of people in Minnesota affected by the outbreak. Sources at the state Capitol told a local reporter that a Schwan's executive had called the governor's office in early October asking that Osterholm be "reigned in or dismissed."[7] However, the company's public affairs manager, David Jennings, a former speaker of the Minnesota House of Representatives, denied that Schwan had tried to get anyone fired.[7]

Hennessy recalls that managing and getting past the behind-the-scenes political pressure was one of Osterholm's many strengths. Shielded from the legal disputes, Hedberg's team continued to take action to protect the public's health and move the investigation along. After the press releases went out, from both Schwan and MDH, asking the public not to eat Schwan's products until further notice, MDH established a call-in line for people who became ill after eating ice cream, and the medical detectives interviewed as many of them as possible. Assisted by MDH statistician Karen White, Hennessy made a list of patients and questionnaire responses and computerized it, a relatively new public health practice in 1994, when few people had personal computers or used the Internet. Although the patients with laboratory-confirmed *Salmonella* included people of all ages from babyhood to age eighty-four, Hennessy determined that the typical patient was a thirteen-year-old boy. On average, the patients were sick for about eight days with fever, chills, vomiting, and diarrhea, but a significant number—about one fourth of the people who contacted the MDH call-in line—were hospitalized.[8] Some had severe diarrhea, dehydration, or complications such as meningitis, septicemia, or Reiter syndrome (a chronic form of inflammatory arthritis).

The medical detectives were anxious to confirm that the case-control study had been correct. Had it really been necessary to dump all that ice cream? Additional evidence came in from several sources. First, a phone survey of Schwan's home-delivery customers strengthened the association between illness and ice cream, indicating that 6.6 percent of ice-cream eaters had experienced fever, chills, and diarrhea. Second, state epidemiologists all over the country, alerted by the CDC, reported *Salmonella* Enteriditis infections among Schwan's customers in their jurisdictions, including forty-eight cases in Wisconsin and fourteen cases in South Dakota, which had reported twenty infections in all of 1993.

As part of the MDH phone survey, customers were asked how many people in the household had shared each gallon of ice cream and whether they still had ice cream in their freezer. If the family had an opened carton of ice cream, the interviewer urged them to throw it out. If the family had an unopened carton of ice cream, the interviewer arranged to collect it for testing. In either case, the interviewer asked the customer to read off the company codes on the ice cream carton. Afterwards, with help from Schwan's, the epidemiologists used the codes to identify the production dates of products associated with illness, which they called "hot lots." Bacteria did indeed grow from unopened ice cream cartons from these hot lots, although it took several days. Of 266 unopened ice cream products, produced on thirty-two different days, eight were positive for *Salmonella* Enteriditis. Those eight products were produced on four days: August 25 and 26 and September 12 and 15. The ice cream made on August 26 was associated with half of the cases of illness identified in the phone survey.

One final piece of evidence convinced everyone that the recall had been necessary. Once Schwan announced the recall, closed its factory, and stopped delivering ice cream, the outbreak ended. Up until then, there had been a continuous rise in cases for more than a month, presumably because people continued to eat contaminated ice cream stored in their freezers. Now there was only a trickle of cases, probably because not everyone heard or paid attention to Schwan's public warnings.[9]

The phone survey allowed the epidemiologists to estimate the size of the outbreak—which turned out to be far bigger than they had imagined, both locally and nationally. First, they multiplied the number of gallons of ice cream distributed in the state of Minnesota (138,000) by the average number of people who shared each gallon (3.2), and then by the percentage of ice-cream eaters who had fallen ill (6.6 percent). By this reckoning, 29,100 people in Minnesota had contracted *Salmonella* from eating Schwan's ice cream. Assuming that the number of illnesses in a given area would correlate with the amount of ice cream distributed in that area—a relationship that held true in all five regions of Minnesota—they estimated that 224,000 people had been affected, nationwide. 224,000 people! The largest common-source outbreak of *Salmonella* ever reported in the United States! The statistical uptick in food poisoning in southeastern Minnesota was indeed just the "tip of the iceberg."

But they still did not know where the bacteria came from, or how the bacteria got into the ice cream. To figure that out, the detectives studied up on how Schwan's ice cream was made. (Hennessy says that's one of the things he likes best about public health—that "you are always learning something new.") The manufacturing process began with ice cream "premix"—a rich blend of milk, sugar, and cream—that the company purchased from two local dairy companies. At the factory, the premix was transferred to storage silos, mixed with chocolate or other flavorings, and frozen into ice cream products. The medical detectives ruled out contamination of the milk, sugar, or cream (the raw ingredients), because the premix was pasteurized before shipping. Contamination of the products after leaving the Schwan factory was also unlikely, because the hot lots included more than one product made on the same date. That left the steps in-between: transport of the premix to the plant, storage, mixing, freezing, and packaging.

Hennessy had an opportunity to witness the ice-cream-making machinery first hand, during a visit to the Marshall plant arranged by inspectors from MDA and the FDA. Hedberg was unable to accompany him, because his second child had just been born. Osterholm, who was also invited, asked Danila to go in his stead because he thought his presence might be a source of controversy.

So Hennessy and Danila flew down to Marshall—a prairie town and county seat in southwestern Minnesota, near South Dakota—with the MDA Assistant Commissioner (who was a lawyer) and an MDA inspector who specialized in dairy products. The ice cream factory, which employed 2,500 people in a town of 13,000, was large and impressive. The first thing Danila saw was a sign on the door that said "No cameras allowed." With a camera slung around his neck, he was not altogether surprised when the man who opened the door refused to let him in. However, after a brief argument between the doorkeeper and the Assistant Commissioner, who threatened to bring a court order, they were allowed to enter.

The factory was shut down and silent, pending the results of the investigation. Several workers were nonetheless present, wearing hair nets and white lab coats, as if it were a normal day. Hennessy and Danila met with David Jennings—the Schwan's public affairs manager—and an FDA inspector who was an expert in dairy plants. They also met the plant manager, the plant's microbiologist, and the factory foreman, who gave them a tour of the facility. Everyone had to follow intensive decontamination procedures, which included walking through a pool of disinfection liquid to clean their shoes and wearing lab coats, head nets, and booties.

The factory was very clean, and the air was filtered. The first few rooms contained stainless steel pipes that could have belonged to any type of modern plant. Later they came to specialized rooms with vats full of chocolate chips and fudge, and then a room where the ice cream premix was off-loaded from tanker-trailer trucks into storage containers that looked like the silos on farms. These rooms smelled delicious.

The visitors stopped in a conference room to talk. Hennessy and Danila sat on one side of a long table with the other government officials, and Jennings sat with the factory people on the other side. He was a forceful and imposing man. When the plant manager offered sandwiches, the MDA official explained that government employees could not have lunch unless they paid for it. The plant manager explained he had never had a guest pay for lunch before, but accepted. (The cost was only $7.30 per person.) Everyone at the table was a little tense. When it was time for dessert, Jennings passed around a silver bowl filled with ice cream sandwiches, and each Schwan employee took one. But when the bowl reached the other side of the table, the MDA inspector hesitated. The Schwan people began to laugh, and Jennings explained that the ice cream sandwiches were made in a different factory, not at the Marshall plant! So the visitors relaxed and ate.

After lunch, the FDA and MDA officials made an official inspection of the factory. They checked the loading docks, looked inside the pipes, examined air filtration systems, and disassembled machinery, looking for possible sites of bacterial contamination. Everywhere they went they took environmental swabs to take back to their laboratories.

Meanwhile, Hennessy and Danila were free to wander around the ice cream plant and ask questions. The factory workers were friendly and helpful, describing how they mixed flavorings into huge batches of premix to make chocolate, chocolate chip, and French vanilla ice cream. They explained that each delicious product could be traced back to the freezer where it had been stored, the flavor vat where it had been prepared, and the premix loads that provided its raw ingredients. They loved their work—and were horrified about the outbreak. Some were nearly in tears when they talked about it. Our great ice cream! It couldn't possibly make people sick! They were proud of their beautiful factory and could not believe that contamination had occurred on their watch. Maybe it was the trucks that brought the ice cream premix to the factory.

Back in St. Paul, Hennessy and Danila decided to explore this idea further, especially since the FDA and MDA inspectors had given the factory a clean bill of health. They learned that the premix came from two suppliers, one in Rochester (one and a half hours away) and one in White Bear Lake (three hours away). When not carrying premix, the tanker-trailers carried loads of oils, molasses, corn syrup or other products.

With a little legwork, the epidemiologists turned up a new lead. Until recently, the trucks that served the supplier in White Bear Lake had driven back empty after delivering loads of premix to the ice cream factory in Marshall. However, a few months earlier the trucking company had signed a new contract that required the trucks to continue south to pick up liquid eggs from egg-breaking plants in Iowa and Nebraska and haul them back to Minnesota for processing at an egg-pasteurizing facility in Gaylord, Minnesota. (Liquid eggs—raw egg materials (whites, yolks, or both) removed from the shell—are a popular refrigerated convenience item for home cooks and bakers, as well as a bulk ingredient used by bakeries, restaurants, and food companies.) After July 1, back-hauling a load of liquid eggs before picking up another load of premix had become a common practice.

The medical detectives were intrigued: Could the contamination, after all, be traced back to eggs, the most common carrier of *Salmonella* bacteria? This idea felt right, though it might be difficult to prove. Each truck-load of liquid eggs was made from eggs from many different flocks, possibly on different farms, so it was possible that some eggs had come from a flock infected with *Salmonella*. Although liquid eggs are sometimes pasteurized before leaving a particular egg-breaking factory, in this case the eggs were not pasteurized until arrival at the facility in Gaylord. And although the trucks' tanks were supposed to be washed and sanitized between trips, that did not always happen. Also, scraping egg off surfaces (even in one's kitchen) can be difficult. Maybe contaminated residue from the liquid eggs in the tanker trucks had come in contact with the ice cream premix.

To test this theory—that *Salmonella* was spread via tanker trucks that also shipped liquid eggs—the detectives performed another epidemiologic study. However, this one involved ice cream products and tanker trucks instead of human beings. Hennessy and Danila visited the MDA dairy expert who had accompanied them down to Marshall. He laid out on his conference table many months' worth of shipping records, covering the schedules of eighty-nine tankers that had hauled premix since July 1. Together, the three of them scrutinized the travel logs to figure out which trucks had carried premix that ended up in products made from hot lots of ice cream.

This was not a simple task. Each ice cream product was made from two to nineteen loads of premix, so it was not possible to trace a particular product to a single tainted load. Instead, they worked backward from each hot lot date. For the August 25 hot lot, for example, they examined the logs from August 24, August 23, August 22, and so on. Here is what they found:

- Three of the four tanker-trailers that delivered premix used in the August 5 and August 26 hot lots had carried liquid eggs immediately before carrying premix.
- The higher the percentage of premix loads in an ice cream product that were hauled by a tanker after hauling an egg load, the more likely the ice cream was to be associated with illness.
- There was no correlation between hot-lot ice cream products— or control products matched for flavor and size—and any of the eighteen mixing vats, ten freezers, or ten storage silos at the Marshall plant.

Thus, the only production step that correlated with *Salmonella* contamination was hauling premix in tanker trucks that had also hauled liquid eggs.[8] So here was the point of contamination, a seemingly minor slip-up in sanitary practices involving inadequate cleaning of a few delivery trucks.

Danila decided to visit the trucking company to take a look for himself. He climbed inside a tank and took some pictures. Although the tank was made of bright, smooth stainless steel, he could feel rough edges and fissures near the welded seams. The valve where the eggs and premix were poured in and out had a rubber gasket that could be removed and cleaned between hauls, and the top of the tank had a built-in, high-pressure shower head. The washing apparatus might be good, but not necessarily good enough to remove gooey egg material from every crack and seam.

Danila's observations were in good accord with the official inspection of the trucks conducted by the FDA and MDA. The inspectors found soiled outlet-valve gaskets, inadequate record keeping, and lack of routine inspection of the interior of the trailer. The documentation of cleaning was absent for seven tankers on several occasions, and egg residue was discovered in one of the tankers after cleaning. Although the inspectors found cracks in the linings of five tankers (including a tanker associated with four illness-associated products), no bacteria grew out of their environmental swabs, probably because weeks had gone by and the bacteria were all dead. As the medical detectives like to say, it was a human bioassay—the outbreak of food poisoning—that revealed the presence of *Salmonella* in the tanker trucks. (Later on, however, the FDA found some confirmatory lab evidence. Liquid egg samples obtained from egg-breaking facilities served by the trucking company contained phage subtypes of *Salmonella* Enteriditis that matched those that infected the patients in Minnesota.[8])

Now Hedberg's team had all the information needed to trace the bacteria's route from "farm to table"—as the FDA saying goes.[10] What struck them as most surprising was how few bacteria had made the journey. According to the FDA and MDA laboratories, the amount of bacteria in the contaminated products was vanishingly small—about 0.2 to 6 bacteria per half-cup serving of ice cream![8] It didn't seem possible. How had such a tiny dose of bacteria—less than the amount detectable by the factory's quality-control tests—made two hundred thousand people sick?

The medical detectives believe that the bacteria most likely came from a chicken farm—probably a large egg-laying operation—that experienced an undetected outbreak of *Salmonella* Enteriditis during the summer of 1994. Because *Salmonella* Enteriditis can spread from hen to hen without causing signs of illness, the egg producer who sold the tainted eggs to the egg-breaking factory was probably unaware that bacteria had invaded the ovaries of healthy-looking hens and multiplied inside their eggs before the shells were formed. Very likely the egg producer routinely used these sales to dispose of eggs that could not be sold as grade A shell eggs. At the factory, the yolks and egg whites of the *Salmonella*-infected eggs were spun out and pooled with material from thousands of other eggs, creating tanker loads of contaminated liquid eggs. A single mass production poultry farm that supplied a few hundred eggs per haul could have been responsible for all of the contaminated egg loads.

A tanker-trailer truck is like a huge, air-insulated thermos bottle that is laid on its side and placed on top of an eighteen wheeler. Since the liquid eggs were kept cold within the tanker, the bacteria were probably too chilled to flourish and multiply. But they did not die. When the eggs were poured out of the tanker, some bacteria remained alive in the gunky egg residue that stuck to the seams and the rubber gaskets. A few of these hardy organisms also survived the high-power wash (if one was performed) and were still viable a day or two later when the tanker-trailer picked up a load of premix for delivery to the Marshall plant. At that point, the bacteria could easily have slipped inside the premix. When the truck arrived at the ice cream plant, the tainted premix was pumped into chilled storage silos, where it mixed with other, non-contaminated loads.

This scenario probably recurred several times, so that the contents of silos emptied on four different days—August 25 and 26, and September 12 and 15—were used to prepare tainted ice cream products of different flavors and types. Each of the production steps in the ice-cream-making process—pouring, mixing, packaging—was performed at freezing temperatures, preserving the bacteria but not allowing them to grow. (As Osterholm, says, the bacteria were "frozen in time" inside the ice cream.) Then the ice cream products were delivered to homes, schools, nursing homes, cafeterias, and restaurants, and consumed by people all over the country.

Ingesting a few *Salmonella* bacteria does not usually make a person sick, because most of us have robust immune systems. Typically, we get sick only when the infectious dose is high (thousands of bacteria per bite), so that some bacteria survive the digestive process and grow inside us and cause gastrointestinal illness. When the dose is low, most bacteria perish in the stomach, destroyed by gastric acid, and those that reach the intestines are finished off by immune mechanisms involving antibodies and white blood cells (including "helper" and "killer" T cells) before they can multiply into a mass that is large enough to do harm. In this case, the high fat content of the ice cream may have protected the bacteria from gastric acid, allowing more of them to reach the intestines. But even so, most people did not get sick. In fact, on average only 6.6 ice-cream-eaters out of every hundred suffered from food poisoning.

The reason that the outbreak was so big, in spite of all this, was that the tainted ice cream was consumed by millions of people. The infectious dose and the attack rate (6.6 percent) were low, but the number of people exposed was very high. It is also worth noting that the people who ate the ice cream probably included a significant subset who were especially vulnerable to infection. Individuals who are at special risk include not only elderly people, infants, and pregnant women, but also people living with HIV/AIDS (whose disease attacks the immune system); people with cancer, diabetes, and sickle cell anemia; organ transplant recipients who take immunosuppressive drugs to prevent transplant rejection; and people who take immunosuppressive drugs to treat autoimmune disorders such as rheumatoid arthritis and inflammatory bowel disease. In these patients, *Salmonella* can spread unchecked from the intestines to the blood stream, and then to other body sites, and can cause death unless the person is promptly treated with antibiotics.

In 1992, an influential report from the U.S. Institute of Medicine argued that many aspects of modern life help infectious diseases emerge and spread, especially in industrialized countries.[11] The 1994 ice cream outbreak illustrates this idea very well, because the bacteria's farm-to-table journey was aided and abetted by modern innovations in agriculture, food processing, and medicine. First, a mass production chicken farm provided ideal conditions for silent spread of *Salmonella,* leading to contamination of eggs sold to a facility that produced liquid eggs—a modern agricultural product thought to be safer for home-use than shelled eggs. (In fact, concern about *Salmonella* in eggs and chicken has helped drive the rapid growth of the liquid egg industry.) Second, contamination of processed foods—though very rare—can affect huge numbers of people. In this case, centralized food processing techniques allowed the Marshall factory to produce hundreds of thousands of gallons of tainted ice cream and deliver it to towns and cities in forty-eight U.S. states. Finally, due in large part to advances in modern medicine, such as organ transplantation and immunosuppressive drugs, about ten million Americans (3.6 percent of the population) have weakened immune systems. Hence, the tainted ice cream was enjoyed by people who were particularly susceptible to bacterial infection.

Thus, salmonellosis may be considered a "disease of civilization."[12] Outbreaks of *Salmonella* Enteriditis have become the single most common cause of bacterial foodborne disease in industrialized countries, despite remaining relatively rare in developing countries, where most food is produced and consumed locally. In countries like the United States, the old scenario in which a small group of people ingests a large dose of *Salmonella* in a home-made tuna fish salad at a church picnic or Jesse James Day event has been replaced by one involving low-level contamination of a product made on an industrialized scale and distributed to many U.S. states or even many countries.

Outbreaks of this type are difficult to detect, because *Salmonella* cases in New Jersey do not appear related to cases in Minnesota or Oregon—and if the outbreak is not detected, the outbreak is not controlled. In this case, the recognition that the cases were related was made only because Minnesota's centralized disease surveillance mechanism was sufficiently sensitive to pick up a small signal in a corner of the state with a very high density of Schwan's customers—and also because MDH had sufficient public health resources to pursue the investigation. If the outbreak had not been identified—or if Schwan's had refused to recall the product without laboratory proof of contamination—the tainted ice cream would have remained in people's freezers, and many more people would have fallen ill.

Once the investigation was over, Hennessy and his colleagues wrote up their findings and published them in the *New England Journal of Medicine*.[8] They described how ice cream was implicated as the source of the outbreak by a small but statistically significant study—a satisfying demonstration of the power of epidemiology and statistics. Schwan's voluntary recall of its ice cream on the basis of this study, in the absence of microbiologic data, set a new precedent for rapid removal of contaminated products from the marketplace. Today, we have new laboratory techniques that do not require waiting for bacteria to grow on petri dishes. But, as Osterholm says, great work was done before that, just as great police work was accomplished before DNA testing became available.

Hennessy's article discussed the need for better public health surveillance and more-sensitive techniques for testing foods in the laboratory. It also suggested that food companies systematically review their manufacturing processes to identify production steps that could allow bacterial contamination. Finally, it recommended that premix and other dairy products be transported in dedicated trucks or re-pasteurized after transportation.

The Schwan Food Company took these lessons to heart, striving to regain its good reputation—especially the reputation of its ice cream. When the Marshall factory reopened in November 1994, Schwan leased a fleet of tanker trailer trucks that hauled only pre-mix and began building its own pasteurization facility. To identify and eliminate other potential points of contamination, Schwan's set up an FDA-approved Hazard Analysis Critical Control Point program—a program created in the 1960s to help Pillsbury provide astronauts with nutritious food that would not upset their stomachs during space missions.

Although the company was cleared of direct responsibility for the outbreak, Schwan's settled a class-action lawsuit in 1995 by paying from $80 to $75,000 (depending on degree of illness) to thousands of customers. The company also provided $160 in cash or gift certificates to thousands of others who agreed not to sue. By 1996, the company's sales had returned to pre-outbreak levels. The company had conveyed concern about its customers and instituted new food safety practices. Moreover, its ongoing cooperation with MDH demonstrated how private/public partnerships can protect the public's health.

Over the following years, the Minnesota Department of Health continued to make foodborne diseases a priority. In 1995 MDH became a founding member of FoodNet, a national surveillance system supported by the CDC to track *Salmonella* and other foodborne pathogens.[13] In 1996, MDH helped the CDC launch PulseNet, a national network of public health laboratories that uses modern molecular fingerprinting techniques to link geographically dispersed cases of disease that are actually part of a single outbreak. In recent years, PulseNet has "gone global" to meet the public health challenges of growing international trade.[14]

In 1999, Hedberg left MDH for the University of Minnesota, where he studies foodborne diseases and teaches students to be medical detectives. He says that being a professor has allowed him to enjoy epidemiology while spending more time with his family. Danila and

Hennessy have continued to work at MDH and the CDC, respectively. As the Assistant State Epidemiologist of Minnesota, Danila has investigated many more unusual health events, including the outbreak described in Chapter 6.

Osterholm remained as Minnesota's medical-detective-in-chief until 1999. Over the next decade, he carved out a unique occupational niche as a public health entrepreneur, founding the Center for Infections Disease Policy (CIDRAP) at the University of Minnesota, as well as an Internet-based information company that focuses on infectious disease issues. He also wrote a best-selling book on bioterrorism, advised the Secretary of Health and Human Services on public health preparedness, and appeared on *Oprah* and other TV shows as a nationally recognized expert on food safety, biosecurity, and pandemics. As a Distinguished Professor at the University of Minnesota School of Public Health and Director of CIDRAP, he continues to contribute to national policy discussions about emergency preparedness, influenza vaccines, and whether (and how) to limit biomedical research that might be used for harmful purposes.

Hennessy, who currently heads the CDC Arctic Investigations Program in Anchorage, Alaska, looks back on the 1994 ice cream outbreak as a turning point in his career. Although he loves science and statistics—and everything else about being a medical detective—when he arrived in Minnesota he was not sure he had made the right decision in leaving clinical medicine to join the Epidemic Intelligence Service. His work on the Navajo Reservation had improved people's lives, every single day. Now, however, having experienced two major outbreaks— one as a doctor (see Chapter 7) and one as a medical detective—he was satisfied that public health was a good choice too. The investigation of the tainted ice cream had been compelling—consuming his life, taking up every waking hour for days and days—and its resolution had a powerful and far-reaching impact, not only preventing additional illness but also changing food safety practices and creating new knowledge that would help safeguard people's health for years to come.

6

The Red Mist

In September 2007, an attending physician at the Austin Medical Center (AMC) in Austin, Minnesota, examined a patient with numbness, tingling, fatigue, and weakness in the legs and feet—symptoms that suggested damage to the peripheral nervous system, the network of nerve cells that transmit information between the brain and spinal cord (the central nervous system) and other parts of the body. Through an interpreter, the doctor learned that the patient, an otherwise healthy and physically fit Hispanic immigrant, worked at Quality Pork Processers (QPP), a local meatpacking plant.

The Spanish interpreter was surprised. She had interviewed other patients at the hospital with tingling, pain, and difficulty walking—and she was pretty sure some of them worked at the same factory.[1] Sure enough, a review of medical records by the nursing staff turned up ten cases involving muscle weakness, pain, or tingling going back to December 2006. How had they missed this? Taken individually, the patients' complaints were varied and fairly vague, but taken together—with QPP as a common link—the cluster could be evidence of a disease outbreak.

But an outbreak of nerve damage? That was odd. The attending physician called the local field office of the Minnesota Department of Health (MDH), which in turn called MDH headquarters in St. Paul. Muscle pain and weakness are common in slaughterhouse workers who are on their feet all day, doing the same physical task over and over. The MDH occupational health office thought the patients might be suffering from repetitive motion injuries caused by muscle strain.

Two weeks later, a medical epidemiologist at the Mayo Clinic—which owns the Austin Medical Center—called the state health department with more details. He spoke with Richard Danila, one of the MDH medical detectives who solved the mystery described in Chapter 5. The problem was not muscular, but neurologic. The patients reported tingling and numbness, not just pain and weakness, and their reflexes were abnormal. Could it be an inflammatory nerve disorder—neuropathy—caused by infection? Later the same day, Danila's office was contacted by the chief occupational nurse at QPP, who had referred three workers complaining of tingling, pain, and "heavy legs"[2] to the Mayo Clinic for special testing. She was worried that an epidemic might be spreading through the plant.

Medical detective Aaron DeVries was assigned to the case. He was new to MDH, having joined in July after completing an infectious diseases fellowship and assisting in post-hurricane Katrina recovery efforts as a Medical Reserve Corps volunteer. A minister's son who "grew up all over the country," DeVries is even-tempered, observant, and thorough. Over the course of one long day, he visited the Mayo Clinic in Rochester (about eighty miles from St. Paul) and then the Austin Medical Center and the meatpacking plant, both in Austin (forty miles further on). He learned at the Mayo Clinic that the QPP workers were suffering from polyneuropathy, a neurologic disorder involving damage to multiple peripheral nerves. The symptoms included head and neck pain, muscle weakness, numbness, and pins-and-needles sensations or burning pain in the hands and feet. A few of the patients had lost sensation in their lower bodies and required walkers or wheelchairs.

Polyneuropathies can have many different causes, including an underlying disease like diabetes or physical traumas like sports injuries or car crashes. DeVries could rule out diabetes and traumas as unlikely explanations for a cluster of cases. A more probable cause for a slaughterhouse outbreak was either chemical poisoning or a contagious disease spread from animals to humans. Infectious diseases are, in fact, the most common cause of polyneuropathies.[3]

The pathogen most often passed from hogs to humans is *Trichinella*, a parasitic roundworm ingested in undercooked pork. However,

Trichinella infection causes stomach cramps and diarrhea and is seldom seen in farm-bred pigs, thanks to modern farming practices and USDA rules and inspections.[4] In any case, the QPP outbreak did not appear to involve consumption of contaminated food—a major concern during a food-factory outbreak. The people who worked at QPP, and not the people who ate QPP meat products, were getting sick.

Since the symptoms were neurologic, DeVries also considered the possibilty of a degenerative brain disease caused by a prion—a special kind of protein that causes other proteins in the brain to fold abnormally. Prion diseases have been reported in several mammalian species, though not (so far) in pigs. Examples include mad cow disease in cattle, scrapie in sheep, and chronic wasting disease in deer. Prion diseases that affect humans include Creutzfeldt-Jakob disease (CJD), which causes progressive dementia; a variant form of CJD caused by eating beef from a cow with mad cow disease; and kuru, a fatal neurologic disease identified in the 1950s among New Guinea tribes who consumed the brains of deceased relatives as part of funeral services.[5]

However, prion diseases attack brain tissue rather than peripheral nerves, and they are progressive and incurable. Fortunately, that was not the case here. The first neuropathy patient seen at the Mayo Clinic—later identified as the outbreak's "index patient"—was a young Mexican immigrant who had improved after treatment with steroids, a standard anti-inflammatory treatment, during a two-week hospitalization.[2, 6-8] His illness had begun back in December with flu-like symptoms and muscle pain that he attributed to standing on his feet all day. But when he lost all sensation in the lower half of his body, his family had rushed him to the emergency room. Since then he had recovered twice, but his symptoms relapsed each time he returned to work at QPP.

After completing his consultation with physicians at the Mayo Clinic, DeVries drove over to Austin, where he spoke with the AMC staff and reviewed patient records, looking for additional clues. By this time he was convinced that the disorder was neither food-related nor due to repetitive stress. He had also ruled out an infection transmitted from person to person, because none of the patients' family members were sick.

His next stop was the QPP factory, which is located on the bank of the Cedar River, next to the headquarters of Hormel Foods, the corporation that produces Spam, Dinty Moore stew, and other canned convenience products. Austin, a city of about twenty-five thousand people, is known as "Spam Town USA" because Hormel is its biggest employer. The QPP factory used to be part of Hormel, but was spun off in the 1980s—a time of increased automation and consolidation in the meatpacking industry—after a thirteen month strike that ended with Hormel workers accepting a significant cut in wages.[9] Today, QPP continues to buy hogs from Hormel and provide meat products to Hormel for packaging and sale. Many of its 1,300 employees are Spanish-speaking immigrants from Latin America.

As he entered the factory, DeVries could see the huge holding pen where the hogs, excited and screeching, are unloaded off massive livestock trailers.[10] He met with Kelly Wadding, QPP's owner and chief executive officer, and then with the factory's head nurse, who showed him around the occupational health clinic. She said that some of the workers, whose jobs are physically taxing, were having difficulty climbing the stairs to her third-floor office.

Wadding and the QPP director of human resources gave DeVries a quick tour of the factory so he could see how a modern abattoir operates. In addition to the holding pen, the factory included three large sections: the entry room where the hogs are stunned by electroshock and killed, the "warm room" where the carcasses are butchered, and the "cold room" where the cuts of meat are prepared and packaged.

The warm room is a huge, barn-like area where everything is planned out in great detail. The first step is to hang the hogs by their rear legs, after the blood has been drained and the skin removed. The last step is to put the meat into the freezer for processing in the cold room. Between those steps the hog is disassembled, with each piece removed for packaging as a food product. The pork-processing assembly line is like the assembly line in a car factory, but in reverse (a "dis-assembly line"), starting with a live animal and ending with packaged body parts. As in a car factory, each worker stands at a designated workstation and performs the same task over and over, as each partially processed product comes down the line.

DeVries returned from the tour with a major epidemiologic clue. Kelly Wadding and the human resources manager had made a quick sketch of the warm room, indicating the workstations of the people who had fallen ill. Most were in one small area of the very large plant. DeVries and his boss, Ruth Lynfield, the Minnesota State Epidemiologist, decided to arrange a full-scale inspection of the QPP factory. Whatever the cause of the outbreak—a microbe, a toxin, or something else—they would find it in the warm room.

A New Yorker long re-settled in the Midwest, Lynfield is an articulate, energetic, out going person who knows how to make things happen. She advised DeVries to call James Sejvar, a CDC medical detective who had studied neurology at the Mayo Clinic and graduated from the CDC Epidemic Intelligence Service (EIS). A person trained in both neurology and epidemiology could be a valuable resource. Lynfield and DeVries were pleased to reach him on the first try, because he is in demand and travels a great deal.

Sejvar was familiar with the research work at the Mayo Clinic neurology laboratory. During his medical residency he had met the legendary neurologists Peter J. Dyck and P. James Dyck—father and son—who are experts in peripheral neuropathies. He had also met Daniel Lachance, a research scientist with a special interest in neurologic disorders, in which immune cells or antibodies attack a patient's own nerves as if they were germs or foreign bodies. The close proximity of three such experts was an amazing piece of good luck for the public health investigators. As it happened, Lachance and James Dyck had already examined several ill workers from the pork processing plant before an outbreak was suspected. They thought the index patient's relapsing symptoms suggested an autoimmune condition like chronic inflammatory demyelinating polyneuropathy (CIDP), which involves destruction of the myelin sheath, the protective protein coat that insulates the axon, the long threadlike part of a nerve cell that conducts nerve impulses. CIDP can be triggered by infection, most often by *Campylobacter* bacteria, a common cause of gastrointestinal disease. Here was another reason to suspect that the workers' illness might involve an infection, perhaps an infection acquired from swine.

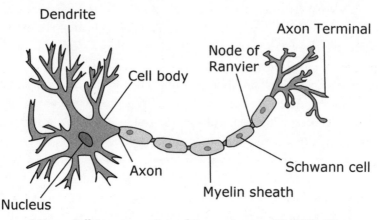

A Nerve Cell. Image courtesy of Quasar Jarosz - CC-BY-SA-3.0.

With this idea in mind, Lynfield asked three animal health experts to attend the factory inspection, as well as an infection control specialist and a toxicologist. She invited a scientist from the Minnesota Department of Agriculture, a veterinarian from the Mayo Clinic with expertise in pathogens that infect both humans and animals, and a veterinarian named Stacy Holzbauer, who was an EIS trainee assigned to MDH. Although most EIS officers are doctors, in recent years EIS has prioritized recruitment of veterinarians, recognizing that most new infectious diseases—including AIDS, severe acute respiratory syndrome (SARS), Ebola hemorrhagic fever, hantavirus pulmonary syndrome, and pandemic influenza—have emerged from animal reservoirs.[11]

Before entering the cavernous warm room, which was large, steamy, and smelly, the inspection team, led by Lynfield and DeVries, donned plastic gloves, booties, masks, and plastic gowns. Wadding supplied a factory map and provided the toxicologist with material safety sheets on chemicals used in the factory. The inspection team walked along the assembly line, observing the tasks performed at each workstation. Cutting up the hog carcasses was a messy and repetitive business. Each worker made a particular cut, over and over, in each successive hog. Some used large knives or other cutting tools, while others used sophisticated machinery. The workers wore different types of protective clothing, depending on their jobs.

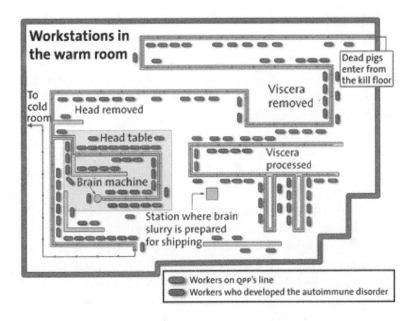

At the very end of the line the inspection team came to the "head table," the place where the hogs' heads were disassembled and their brains removed. About thirty-five workers stood around this table, removing the tongues, cheeks, temples, snout, ears, and jaw of each hog using knives, chisels, and automatic cutting tools. (The factory people liked to say that they utilized "everything but the squeal," so that nothing went to waste.) The next to last job was called "backing heads," chopping and removing muscle from the backs of the skulls with a circular knife, and the last one was "blowing brains," removing the brains from the skulls. These jobs were relatively well-paid and sought-after by the QPP workers, who bid for them on the basis of seniority and ability.[12] A head came down the line about every three seconds, and it required speed, dexterity, and sustained focus to remove as much muscle and brain as possible.

The head muscles and brains of hogs are commercially valuable as specialty food products. Head muscle can be made into sweet or spicy sausage or into headcheese, a meat jelly that can be served with crackers or bread, like goose liver pate. Scrambled eggs with brains is a popular dish in some southern states, where hog brains are sold in supermarket cans as "pork brains in gravy" or bought at meat markets as fishing bait. However, the greatest demand for hog brains is in China and Korea, where they are deep-fried, stir-fired, or cooked in hot pots. From an economic point of view, U.S. hog brain sales to Asia are part of the modern globalization of trade.

The brain-harvesting job was called "blowing brains" because the brain-removal worker "blew" the brains out of the skull with a blast of pressurized air. Standing at the very end of the head table, the brain-removal worker placed each skull, right-side up, lower jaw and skin removed, on the metal hose of a pressurized air device, with the hose fitting through the foramen magnum, the "great hole" where the spinal cord enters the skull. Placement of the skull on the nozzle automatically tripped a wire that caused pressurized air to rush into the skull, liquefying the brain matter into a soggy pink mass that drained into a ten-pound bucket. Another worker periodically carried the pooled brains to the cold room for further processing.

It was remarkable to see the brain-removal procedure in action. The head table worker stuck the skulls on the nozzle as fast as he could, one after the other, generating some splatter, trying hard to keep up. Walking ahead with Kelly Wadding, Ruth Lynfield could see a strange reddish mist blowing around the room, propelled by air currents generated by an overhead fan. (Even though it was November, it was warm in the warm room.) Pink particles settled on the people at the head table, especially on the brain-removal worker and the person backing heads. A Plexiglas shield near the pressurized air device provided little protection, nor did the workers' goggles, coats, and cut-resistant gloves, which left their arms exposed from the shoulder to the wrist. (Few wore apron sleeves, which can be hot and uncomfortable.) Moreover, none of the head table workers wore respirators or face masks, so there was nothing to prevent them from breathing in the aerosolized brain matter, minute by minute, day after day.

Staring spellbound at the head table and its workers—who looked both strong and vulnerable—Lynfield said to Wadding, "Kelly, what do you think is going on?"[13] Wadding responded by deciding, then and there, to stop using air-blasting machinery to harvest brains. On the spot, he ordered the device dismantled and brought to his office. He also agreed to provide the head table workers with additional protective equipment.

Though they did not realize it at the time, the brief exchange between Lynfield and Wadding ended the outbreak of polyneuropathy in Austin, Minnesota. Eliminating the factory's brain mist turned out to be a totally effective prevention measure. Later on, the medical detectives would compare it to "taking the handle off the Broad Street pump"—the iconic action taken by the Englishman John Snow, the "father of epidemiology," to end an 1854 cholera epidemic transmitted through drinking water. In Austin, as in long-ago London, the simple action that ended the outbreak was performed on the basis of a hypothesis. John Snow thought that cholera was caused by a waterborne microbe, and Lynfield thought the polyneuropathy was caused by the brain mist. Neither had proof.

Over the next few days, the medical detectives designed a formal study to confirm or disprove the brain mist theory. At the same time, they considered other possibilities, just in case they were chasing a red herring. One possibility was that the illness was caused by poisoning from a toxic chemical used in cleaning or processing; another was that an airborne toxin or pathogen was spreading through the factory's air vents. However, colleagues from the CDC's National Institute for Occupational Health (NIOSH) ruled out the air-vent theory after reviewing the factory's blueprints and maintenance records. The NIOSH inspectors also confirmed the toxicologist's conclusion that the factory did not use chemicals whose ingestion or inhalation could cause neurologic symptoms.

Lynfield asked Holzbauer, the veterinary epidemiologist, to lead the epidemiologic study, because she was the only member of the MDH team who was familiar with pork processing plants. Holzbauer grew up in South Dakota as the eldest of eight children on a "grain, cattle, and kid farm." Her plan was to become a large-animal veterinarian, marry a cowboy, live on a ranch on the Great Plains, and raise cowboys. However, after a few years as a large-animal vet, she enrolled in a master of public health program for veterinarians, inspired in part by Laurie Garrett's public health classic, *The Coming Plague*. She was surprised to discover that public health and large-animal medicine share a basic world view, because epidemiologists (like large animal vets) essentially view their patients as a herd—members of a single species that live together in large groups—rather than as individuals. (Small-animal medicine, caring for individual pets, is more like human medicine.) After working on a CDC project to promote appropriate use of antibiotics for veterinarians, Holzbauer—convinced she had found her niche—applied to the EIS program. Today, she is happily married (though not to a cowboy) and the mother of twin baby girls.

Holzbauer knew all about hogs. Her family raised them when she was in grade school, and she had studied the parts of a hog's lifecycle—birth, growth, finishing, slaughter—in college and vet school. Holzbauer thought that the QPP factory was fairly typical, though she had never before seen compressed air used to harvest hog brains. In fact, when she first saw the compressed-air brain-removal device, she thought it was a good idea—less chance of cutting off fingers with axes or bandsaws!

Holzbauer developed a questionnaire to help the investigators identify disease risk factors by comparing the experiences of the QPP patients and the workers who had remained healthy. Thirteen of fifteen patients identified by AMC, the Mayo Clinic, and the QPP nursing staff enrolled in the study, as well as two groups of healthy workers who served as controls: forty-nine who worked at the head table and fifty-six who worked anywhere in the warm room.[12] All thirteen neuropathy patients had worked in the warm room, and seven had worked at the head table. Eight were women, and 80 percent were Hispanic. Their symptoms included numbness; prickling or tingling in the legs, feet, hands, or arms; decreased strength; and absent or decreased reflexes. Some reported severe fatigue, searing skin pain, leg cramps, or a feeling that the soles of their feet were on fire. The diagnosis of peripheral neuropathy was confirmed by advanced electromagnetic testing at the Mayo Clinic that indicated damage to the nerve axons or myelin sheath.

The patients and the healthy controls were interviewed about their health histories, jobs and work histories, and potential sources of exposure outside the factory. It was requested that each interviewee stop by the factory's medical station to provide a throat swab and blood sample. The study went smoothly, with the plant managers arranging for workers to be taken off the line, three at a time, for forty-five minute interviews. They also provided blueprints so the team could measure the distance between the patients' workstations and the site of brain removal. Lynfield was pleased at the degree of cooperation. She says it is rare that things are done appropriately, with support rather than resistance. She and Wadding held joint press conferences and met with the workers to explain the purpose of the study.

It was especially important to gain the trust and cooperation of the factory workers, including QPP's Hispanic workers. Holzbauer's team provided informational materials in Spanish and arranged for native Spanish speakers to conduct interviews. The interviewees were not required to give their names and were assured that the investigators' mission was to improve the plant's health conditions and not to enforce any laws or regulations. DeVries remembers being at the factory late at night in the middle of winter to talk with night shift workers. Being there was important, because it helped demonstrate that the team was interested in the employees' welfare.

The results were clear: the thirteen neuropathy patients were 6.9 times more likely than the warm-room controls to have ever worked at the head table and 7.5 times more likely than the head-room controls to have either blown brains or backed heads.[12] Blowing brains was the job held by the index patient, whose illness had relapsed twice, and backing brains was the job of a patient who told a *New York Times* reporter that she "always had brains on my arms."[2] These two individuals were the most severely ill. Like the index patient, the woman who backed brains had fallen sick months before the outbreak was detected. Her illness began with tingling in her hands and knots in her legs and progressed over months until she could no longer stand long enough to do her job.[2, 7, 14] She had not realized her illness was due to an outbreak until she saw a report on television.[7]

No correlation was found between illness and prior vaccinations, medications, insecticides, pesticides, contact with other animals, travel history, or activities conducted outside the factory. However, while proximity to the brain-removal station was the strongest predictor of disease, not every neuropathy patient had worked at or near the head table. Danila recalls that one of the patients worked in a different section of the warm room, some distance away from the brain-removal device. The epidemiologists considered her an "outlier" who did not fit their hypothesis. However, it turned out that her best friend worked near the brain harvest station and that she spent her coffee breaks, twenty minutes a day, standing next to her friend as she worked. (She did not sit, because there were no chairs at the head table.)

While the study was in progress, Sejvar and colleagues at the CDC and the USDA surveyed U.S. factories to identify other plants that used compressed-air devices to process hog brains. Of twenty-six federally-inspected swine abattoirs in thirteen states, they found six that processed swine brains, including three that used pressurized air: one in Delphi, Indiana, one in Fremont, Nebraska (owned by Hormel), and the QPP factory in Austin.[12] Sejvar also notified the World Health Organization, the World Organization for Animal Health, and the European Centre for Disease Prevention and Control,[15] but those organizations found no examples of overseas factories that used compressed air to harvest hog brains.

Like the Minnesota factory, the pork processing plants in Indiana and Nebraska had designed their own home-made pressurized-air devices. But the devices in Indiana and Nebraska did not use an automatic trigger. In Indiana, the device had a foot pedal, which provided the operator with more control. In Nebraska, the device was like a gun, completely portable. It had an attachment for a hose like the ones on air pumps for filling tires. Squeezing the attachment let loose the blast of air.

The managers at the Indiana and Nebraska plants did not recall any cases of muscle weakness, numbness, or pain, and no documented cases of neurologic disease were recorded in their occupational health records. However, the medical detectives decided to look further, aware that plant managers and workers might be reluctant to speak with government officials. The Indiana Department of Health and the CDC eventually turned up five cases in the Delphi area, after distributing health-alert notices to local neurologists, healthcare facilities, and county health officials. All five had previously worked at the Indiana factory.[12, 16] A single case of peripheral neuropathy was eventually identified in Nebraska, involving a brain-removal worker who had terminated employment for medical reasons.[12]

In January 2008, a CDC/NIOSH team, including Holzbauer and Sejvar, visited the plant in Indiana to conduct a second risk-factor study, similar to the one completed in Minnesota.[16] The study was led by an EIS officer assigned to the Indiana Department of Health. Holzbauer helped conduct interviews and collect blood samples, and Sejvar conducted on-site assessments of interviewees who reported weakness, tingling, or pain. He identified four additional cases of polyneuropathy among people currently working in the factory, bringing the total to nine. More than half of the illnesses occurred in women, most of whom had been hospitalized, and all of whom were Hispanic.

In Indiana, a worker with a warm-room job called "hanging heads" removed each hog's head from its carcass and carried it to a separate room—called the "head-boning room"—for further processing. The workstations of the ill workers were clustered in the head-boning room, near the brain-removal device. The results of the Indiana study were also clear-cut and statistically significant. All of the patients who participated in the study had harvested brains at least once, and most reported that brain tissue had entered their eyes, nose, or mouth while they worked. Moreover, the amount of exposure time was correlated to risk. Each additional three-month period spent harvesting brains increased the likelihood of illness by 1.5 percent.[16]

The identification of polyneuropathy patients at two other "brain mist" plants was further evidence that hog brain material was associated with the cases of nerve damage. However, one piece of information did not fit: the compressed air devices were not new. They had been in use in Minnesota and Nebraska since 1998 and in Indiana since 1993. Intensive case-finding had not turned up any cases earlier than 2006 in Minnesota or 2004 in Indiana. The single case in Nebraska was identified in 2006.

What had changed? The medical detectives checked the delivery records of the factories in Minnesota, Indiana, and Nebraska but found no common source of hogs during the years preceding the outbreak. The hogs came from farms all over Canada and the United States, mostly from Minnesota, Iowa, Wisconsin, and South Dakota. Moreover, there were no reports of illness at other plants that received hogs from the same farms.

Another possibility was that workers had been falling ill for some time but no outbreak had been detected because of the high turnover at the plant. There was some anecdotal evidence for this, including a patient in Indiana who said she had replaced a worker who got sick while harvesting brains,[17] and an immigrant worker seen at the Mayo Clinic in 2005 who had gone back to Mexico before her tests were complete.[7, 18, 19] It was certainly possible that other workers had left the factory rather than seek help or take medical leave.

Another idea, suggested by the patients themselves,[7, 18] was that the head table workers had breathed in greater amounts of brain mist since 2006 because QPP was slaughtering hogs at a faster rate. The line speed had increased from 1,200 to 1,300 hogs per hour in April 2003, and then to 1,350 hogs per hour in 2006, the year the index patient fell ill. (For comparison, the line speed in 1996, when the device was new, was about 900 hogs per hour.[7]) According to the workers, the need to remove a brain every few seconds made it difficult to place each skull fully on the nozzle, causing misfiring, skull breakage, and greater spray of aerosolized particles. More material escaped into the air, so that the workers stood in a constantly replenished haze of the reddish-grey brain mist observed by Lynfield and her colleagues.

The Nebraska and Indiana factories also reported production increases that fit the outbreak's timeline. The line-speed in Nebraska increased from 1,200 to 1,250 hogs in 2006, the year that a brain-removal worker became too ill to work. At the Indiana plant the situation was a little different because not all hog heads were harvested for brain. Instead, the percentage varied from year to year, based on market demand. However, the Indiana plant had increased its brain harvest rates by 250 percent between 2004 and 2007, and the neuropathy cases were reported between April 2005 and December 2007.[16]

QPP apparently sped up the assembly line because of increased demand for cheap meat products like Spam as the recession took hold.

Instead of hiring additional workers, QPP encouraged overtime work that increased the exposure of individuals who worked at the head table. In Nebraska, the increased demand for brains (as opposed to hog meat in general) may have reflected the decreased availability of cow brains after 2004. This was due to a USDA rule that banned the sale of bovine brains from older cattle because of concerns about mad cow disease.[16]

So far, so good. But the question of how the hog brain material had caused the workers' illness was still unanswered: Was an infectious agent present in the brain mist? Lynfield, DeVries, Holzbauer, and Danila returned to the factory to retrieve samples of not-yet-sold hog brains stored in QPP's cold room. Hog pathogens are well studied, because of their impact on the agricultural industry. A few, like hepatitis E, are known to infect humans; several cause neurologic symptoms (e.g., encephalomyocarditis virus and pseudorabies virus) or embryonic and fetal death (e.g., porcine parvovirus); and others are the porcine relatives of human viruses that cause such diseases as influenza. Despite an exhaustive search, however, no bacterium, virus, or parasite was found in either the hog brain samples or the human throat swab samples collected during the risk-factor study. To leave no stone unturned, the detectives also arranged for testing of stool samples from four Minnesota and two Indiana workers as part of the CDC Unexplained Diarrhea project, knowing that CIDP can be triggered by *Campylobacter* bacteria, a common cause of gastrointestinal symptoms. However, no consistent evidence of a diarrheal infection was found.[12]

But what if the brain mist contained a rare or unknown pathogen for which they had no test? Infection of slaughterhouse workers with unusual animal-borne pathogens is not unheard of. In 1999, for example, a previously unknown virus called Nipah—first identified on pig farms in Malaysia—caused an outbreak of respiratory and neurologic disease among abattoir workers in Singapore who slaughtered pigs from the affected Malaysian farms.[20, 21] Moreover, Q fever—an animal-borne disease described in Chapter 4—has caused outbreaks in sheep or goat abattoirs in France,[22] Australia,[23] Scotland,[24] the Netherlands,[25] and California.[26] However, additional testing, including a search for rare or unusual microbial ribosomal RNA sequences, yielded only negative results.

There was one more possibility: An infection that triggered an immune disorder could have been over before the throat swabs were taken. In that case, the causative microbe might be gone, but microbe-specific antibodies might remain in a patient's blood. This explanation, too, was ruled out, after extensive laboratory testing.

Having found no evidence of infection—either as a direct cause of the neurologic symptoms or as a precipitating factor—the medical detectives began to entertain a more unusual hypothesis that initially struck them as bizarre. Could the brain mist itself—rather than a toxin or pathogen carried along with it—elicit a human autoimmune reaction, similar to a very severe allergic reaction?

The evidence for an autoimmune response was solid, despite the lack of an infectious trigger. The Mayo Clinic doctors had suspected an autoimmune disease from the beginning, because of the pattern of destruction of the peripheral nerves. When they measured the velocity of nerve impulses along the patients' axons, they detected damage at the nerve roots (where the nerves emerge from the spinal cord) and the extremities (where the nerves connect with the muscles or sensory organs), with the mid-nerve sections remaining unharmed. This finding is consistent with immune-mediated nerve damage, because antibodies have greater access to the ends of the nerves, which are well supplied with blood vessels. Damage to the nerve ends was confirmed by several other tests, including magnetic resonance imaging (MRI), electron micros-copy, and nerve biopsies. For example, the MRI found inflammation (irritation and swelling) of the nerve roots. An autoimmune condition was also suggested by the patients' improving when no longer exposed to the brain mist or when given steroids or other immunotherapies.

Additional data implicating an autoimmune disease was pro-vided by Ian Lipkin, the molecular biologist who contributed to the investigation of the New York City outbreak described in Chapter 1. When Lipkin tested the patients' blood serum samples for cytokines and chemokines—proteins released by immune system cells to acti-vate and sustain an immune response—he found high levels of interferon-gamma (IFN-y), a cytokine associated with autoimmune disorders.[12] The Mayo Clinic neurologists confirmed this result and identified many other markers of autoimmune diseases, including high levels of protein in the patients' cerebrospinal fluid.[27]

The idea that the brain mist itself had caused an autoimmune response began to seem less bizarre and more likely. As Sherlock Holmes told Watson, when you have eliminated the impossible, whatever remains, however improbable, must be the truth.[28] A literature search turned up a few, mostly long-ago, examples of neurological tissue from one species causing nerve damage in another. For instance, Louis Pasteur's first rabies vaccine, made in 1895 from virus extracted from canine brain and spinal tissue, had caused neurologic symptoms in some patients.[29] Moreover, in 1935 scientists developing an animal model of acute neurologic disease induced encephalomyelitis (inflammation of the brain and spinal cord) in monkeys by injecting them with liquefied rabbit brain tissue.[30] In the 1980s, encephalomyelitis was reported in patients who received a rabies vaccine contaminated with myelin basic protein, which is a major building block of the myelin sheath.[31] More recently, in the 1990s, an experimental treatment involving injection of bovine tissue contaminated with brain proteins induced Guillain-Barré syndrome (GBS)—an acute form of CIDP—in some patients.[32]

Our modern understanding of molecular biology suggests that Pasteur's observations and some other neurologic autoimmune events might be due to "molecular mimicry," a phenomenon in which antibodies recognize part of a foreign body—an "epitope" or "antigenic determinant"—that is similar to part of a normal human protein.[33] Bovine brain tissue, for example, could have caused GBS in humans by eliciting antibodies that bind not only to cow-brain epitopes but also to human epitopes with a similar molecular structure. Because the human epitopes "mimic" the bovine ones, the human immune system cannot distinguish between them and attacks them both. Similarly, inhalation of aerosolized hog proteins by the QPP workers could have generated "self-antibodies" that attacked both the hog material and the patients' own nerves, causing the sustained inflammation that characterizes autoimmune disease. This explanation is plausible because in fact hogs and human share many proteins in our nervous systems.

Lachance and his colleagues tested this idea by examining the pattern of neuronal self-antibodies generated by the patients from Minnesota and Indiana. The scientists used a well-established laboratory assay developed by a Mayo Clinic colleague, Vanda Lennon,[18] to detect antibodies generated in cancer patients against both cancer cells and normal nerve cells. The assay uses a fluorescent stain to detect human antibodies that bind to mouse neural tissue immobilized on a slide. The result was surprising: antibodies from all of the neuropathy patients—as well as one third of head table workers who had no symptoms—bound to mouse nerve structures, including the myelin sheath, in a pattern the researchers had never seen before, despite having tested thousands of patients with cancer, CIDP, or GBS.[27, 34, 35] They regarded this pattern as the unique biological signature—or biomarker—of a new autoimmune disease.

The new disease has variously been called progressive inflammatory neuropathy (PIN)[36, 37] or occupational inflammatory polyradiculoneuropathy.[38] From a clinical and public health point of view, its identification and characterization marked the end of the outbreak investigation. The source of the outbreak was known, and the patients had been identified and treated with immunotherapies. Above all, the John-Snow-like intervention had worked! No more factory workers were afflicted with the alarming combination of pain, tingling, and loss of sensation. Moreover, Sejvar and his USDA colleagues had alerted the swine industry and abattoirs across the nation and around the world about the dangers associated with the air pressure method of hog-brain removal.

Nevertheless, many scientific questions remained unanswered. A group of Mayo researchers, including Jeffrey Meeusen, Vanda Lennon, Lachance, and Peter Dyck, decided to investigate the mechanism of disease by creating an animal model of "breathing brain."[39, 40] They smeared liquefied hog brain tissue on the noses of anesthetized mice and took periodic blood samples to check for the generation of self-antibodies, which increased in a dose-dependent manner. When the mice inhaled hog-brain material twice a day, five times a week—mimicking the occupational exposure of the QPP head table workers—their immune systems generated the same biomarker pattern of self-antibodies as the QPP workers. Although the scientists could not tell whether the mice experienced tingling sensations or numbness, MRI testing revealed the same swelling of the nerve roots observed in the human patients.

The next step was to determine which nerve structures were harmed by the autoimmune response. Using standard immunologic tests, the neurologists detected self-antibodies to a variety of neural proteins in the blood of the exposed mice, including myelin basic protein and the channel proteins that create membrane pores that allow ions to flow in and out, generating the nerve impulses that travel along a nerve's axon. Which were most significant? Two pieces of evidence—one achieved through careful measurement and one through serendipity—pointed to voltage-gated potassium channels (VGPC) as the antibodies' principal target. First, quantitative testing indicated that the highest levels of antibodies were directed against VGPC protein complexes. In addition, a lab technician noticed that the mice began to shake when administered an anesthetic called isoflurane that works by interfering with potassium channels. The neurologists remembered that a shaking behavior related to isoflurane had been reported before, in the literature on *Drosophila* fruit flies, a major experimental organism for molecular geneticists. In fact, the first gene encoding a potassium channel protein was cloned in 1987 from a *Drosophila* fruit fly mutant called "shaker" because it shook its legs vigorously when exposed to isoflurane.[41] Thus, the self-antibodies may be binding to a VGPC epitope that is common to swine, humans, and possibly also fruit flies.

The neurologists concluded that the factory workers' disease was a channelopathy—a neurologic disease caused by disruption of nerve impulses involving membrane channels that allow passage of calcium, sodium, or potassium ions. Other channelopathies include cystic fibrosis, myasthenia gravis, and some seizures and migraine headaches. These diseases may be caused by inborn mutations or by an autoimmune reaction triggered by a foreign body. In the case of the Austin outbreak, the antibody attack on the potassium channels had led to weakness, pain, and (in some cases) lower-limb paralysis among a group of hard-working people whose jobs required substantial strength and stamina. Although motor function improved in all of the treated slaughterhouse workers,[27, 42] a year and a half later many continued to experience headaches or pain in their necks, lower backs, or feet, apparently due to stretched nerve roots and irritated nerve terminals.[43] Several required medical accommodations, such as sitting for fifteen minutes every two hours, in order to return to work.

The QPP patients were legally entitled to workers' compensation benefits, if they agreed not to sue. (Minnesota is one of nearly two dozen states that allow undocumented workers to be covered under workers' compensation laws.) Twelve eventually received settlements of about half a year's pay ($12,500), and the index patient, who had suffered permanent nerve damage, received $38,600.[7] QPP apparently made some attempts to accommodate the ill workers by providing them with lighter duties. Nevertheless, some workers who filed Workers Compensation claims were eventually fired for working under forged or stolen identities,[7, 44] and the index patient's settlement required him to leave the factory for good.[7] Few left voluntarily, because the QPP jobs, though low-paying and numbingly repetitive, provided a steady livelihood, with health benefits and overtime pay.

Looking back, the medical detectives recall the surprising complexity of the investigation and the intensity of working on a severe and debilitating neurologic disease that had not been identified before. The MDH team worked with experts from academia, industry, and government who examined the issues from many different angles, including animal health, environmental health, neurology, toxicology, immunology, food safety, and agriculture. Holzbauer and DeVries—the junior members of the team—gained invaluable experience in solving a medical mystery through collaborative investigation, and Lynfield and Danila gained new insights into occupational health by by seeing factory workers process hogs at a rate of nineteen thousand a day. In regard to the scientific findings, Sejvar notes that the Mayo Clinic mouse model has already advanced our understanding of autoimmunity by demonstrating that self-antibodies are not just a biomarker but actually play a direct role in causing disease. He is continuing to work with colleagues at the University of Minnesota to learn more about the binding of anti-hog-brain antibodies to human neural tissue.

Above all, however, the medical detectives recall the amazing John Snow moment that stopped the outbreak in its tracks—a moment that all epidemiologists imagine, but few experience.

7

A Normal Spring

Dr. Thomas Hennessy remembers the spring of 1993 as green, lush, and beautiful, in the special way of desert lands after rain. The branches of the piñon trees were laden with nuts, and his pet cats were catching lots of mice and bringing them into the house and garage. The neighbors complained a bit about the mice, but otherwise it was a normal spring in Crownpoint, New Mexico, a small town on the eastern edge of the Navajo reservation. The flu season was over, and the clinic of the hospital where he was chief of staff was treating more patients with allergies and fewer with respiratory diseases.

Hennessy had met his wife, Deanie Golnick, at a residency program at the University of Minnesota in Duluth that trains family-care physicians to work in rural areas, delivering babies and performing basic surgery without back-up from a specialist. Now they were living in a small wooden house a block from the Crownpoint hospital, a twenty bed Indian Health Service (IHS) facility with a clinic and emergency room but no intensive care unit (ICU). Their hospital served more than twenty-five thousand people, scattered through a very large but sparsely populated area. Many were Navajo farmers or herders who maintained a traditional lifestyle, living in dome-shaped, one-room hogans without electricity or piped-in water. They came to the hospital only when they were extremely sick or severely injured. Anyone who required intensive care or complex surgery was transported to a larger IHS hospital in the town of Gallup, an hour's drive from Crownpoint, over a mountain pass and along dirt roads. The nearest big-city hospital was in Albuquerque, 120 miles away.

By 1993, Hennessy and Golnick had lived in New Mexico for three years. They liked their new home, and the birth of a daughter a year before had drawn them closer to the local community. They had also recruited doctor friends from Duluth to join them at the IHS hospital, where their work kept them busy night and day.

The hospital always had two physicians on call to deliver babies, monitor the wards, and take care of emergencies. As new parents, Hennessy and Golnick split their call assignments. On Saturday night, May 15, Golnick admitted a young woman whom she knew fairly well because she had delivered the woman's baby, her first child, the previous fall. She was twenty-one years old, physically fit and athletic, a long-distance runner who lived in the nearby town of Little Water and managed the track team at the Indian School in Sante Fe.[1] Her flu-like symptoms did not seem especially serious. But because she had a history of asthma—and because she had not improved after consulting a local doctor—Golnick decided to admit her for observation, planning to order an x-ray in the morning if she did not feel better by then. The next morning, Golnick signed the patient over to Hennessy, requesting that he look in on her. Finding the patient still coughing and uncomfortable, Hennessy called in the x-ray technician, who was on call at his home nearby.

Just then a young woman entered the clinic, about to have a baby. In such a small hospital, everything grinds to a halt when a baby is about to arrive, as everyone prepares for the big event. This time, the birth went smoothly, and a baby boy was delivered without complications. The staff was very happy. They realized that it was Mother's Day.

Afterwards, Hennessy went to look at the x-rays of the woman admitted the day before. His feelings of happiness disappeared. The woman's lungs were full of fluid. Her x-rays were like those of ICU patients with acute respiratory distress caused by infection or trauma. He went to the patient's room, where he found her struggling to breathe and coughing up frothy white fluid. The fluid did not contain blood, which would be a sign of an infection like TB. However, her blood work looked normal. She wanted to sit up, to breathe better, and the nurse gave her oxygen, as well as a nebulizer for her asthma.

But she continued to cough up the frothy white liquid—a sign of severe respiratory failure.

Hennessy had never before seen anything like this in a young and previously healthy person. He called the University of New Mexico hospital in Albuquerque and asked for a helicopter to come pick her up. Then he gave her more oxygen. Her blood pressure was dropping, so he also gave her intravenous fluids. When that did not help, he inserted a tube into her airway to keep it open. He threw everything he had at her illness, but nothing he did made any difference. It was an awful, chaotic scene. By the time the helicopter arrived she had gone into cardiac arrest and shock. Hennessy and the other on-call doctor made a full resuscitation effort—chest compressions, mouth-to-mouth resuscitation, defibrillation—but still the young woman died.

How could this be? Hennessy and Golnick knew this woman. She had entered their clinic healthy, with mild respiratory symptoms, and twelve hours later she was dead. In modern healthcare, young people do not suddenly die of pneumonia unless something is very wrong. Was it a case of severe bacterial or viral infection? Was it contagious? Hennessy stumbled through the rest of the day. The next morning he described the case to the hospital staff. He told them about the white, frothy fluid, which forms when intercellular fluid leaks into the alveoli, the tiny air sacs of the lungs, preventing the exchange of oxygen and carbon dioxide.

Hennessy was at home three days later when he received a phone call at 10 PM from a friend and coworker (a recruit from Duluth) who was the physician on call that day. He asked Hennessy to return to the hospital to observe a patient in the emergency room. It was the dead woman's fiancé, a strong, hard-working nineteen-year-old who had attended the Sante Fe Indian School on scholarship and won a state championship in cross-country running.[1, 2] He had a similar set of flu-like symptoms—back pain, achiness, low-grade fever, plus vomiting and diarrhea.[3] He was in a bad emotional state too, having lost his sweetheart and been left alone with an infant son. He was bereft and scared.

The man's lab work looked normal, and there was no immediate reason to keep him at the hospital. Nevertheless, Hennessy and his friend urged him to stay. However, the man was anxious to return home and prepare for his fiancée's funeral. The next day he began to feel worse. On the way to the funeral his condition deteriorated until he was in extreme distress, hardly able to breathe. Despite frantic efforts by an ambulance crew, he died before reaching the emergency room of the Indian Medical Center in Gallup.

The local community was devastated. Bruce Tempest, a senior physician at the Gallup hospital, reported the sudden deaths of the woman and her fiancé to the Office of Medical Investigations, New Mexico's state-wide coroner's office. A few days later, an Office of Medical Investigations official told Mack Sewell, the New Mexico State Epidemiologist, that post-mortem findings on the young couple were similar to those of a thirty-year-old woman who had died three weeks before. The patients' lungs were not inflamed, as typically seen with pneumonia, but were filled with clear, yellow fluid called plasma— the non-cellular portion of the blood, made of water and blood proteins.[4] That same day, James Cheek, Head of Epidemiology at IHS in Albuquerque, reported that physicians at other IHS facilities had identified two additional cases of the fatal respiratory illness, bringing the total to five.

The New Mexico public health laboratory quickly ruled out plague, anthrax, and tularemia—the animal-borne bacterial diseases that are the most common causes of sudden respiratory failure in the southwestern United States. Thinking outside the box, Sewell wondered if a toxic chemical had leaked from Fort Wingate, a munitions depot outside of Gallup. However, an official at the U.S. Department of Defense denied that any spills or accidents had occurred. Another possibility, suggested by Cheek, was accidental inhalation of phosgene, a gas used to kill prairie dogs that carry bubonic plague. But no canisters or phosgene pellets were found in the couple's home.[5] On the phone with C. J. Peters, the head of the CDC Special Pathogens lab— which handles scary viruses like Ebola—Sewell reviewed potential viral causes of acute respiratory distress, including the worst-case scenario: a new strain of influenza like the one that killed thousands of young adults during the flu pandemic of 1918–1919.

By this time it was evident that the mysterious outbreak was not limited to New Mexico. Additional cases were reported throughout Four Corners, the area where New Mexico, Arizona, Colorado, and Utah meet. On May 28, the Friday of Memorial Day weekend, Cheek called Sewell from Crownpoint, where he was interviewing affected families and gathering blood samples. Cheek and Sewell agreed that something big and usual was taking place and it was time to request assistance from the federal government.

That afternoon, Jay Butler, a medical detective in the CDC's bacterial diseases division in Atlanta, heard about the outbreak from a colleague on his way to a briefing in the office of James Hughes, the Director of the National Center for Infectious Diseases. He expected that Hughes would send an Epidemic Intelligence Service trainee (an "EIS officer")—or maybe two—to assist IHS and the New Mexico Department of Health. But that is not what happened. Sewell had requested a senior person, and Hughes picked Butler for the job. Not knowing what had caused the outbreak, he was unsure whether a bacterial expert was the best choice. However, he believed that Butler's calm demeanor and knowledge of bacterial pneumonias might help. Born and raised in North Carolina, Butler is a warm person with a shy smile and jaunty mustache. In his student days he planned to become a "regular doctor" but was drawn to public health by a need to understand why people get sick and how illness can be prevented. His experience caring for AIDS patients in the days before antiretroviral therapy further strengthened his interest in disease prevention.

That night, as Butler prepared to leave, his eldest daughter, eight years old, asked him, "When you are finding out why people are dying how do you keep yourself safe?" He assured her that he would wear gloves and a mask and take good care of himself, but at the back of his mind he was worried, both for himself and his family. At the time, Butler had four children, and the youngest had been born only two months before, in March.

After arriving in Albuquerque, on Saturday, May 29, accompanied by two EIS officers, Butler learned that a teenage girl had collapsed the previous evening at a Friday-night dance in a state park outside of

Gallup. She was carried to the hospital where a full resuscitation was attempted, but failed. Butler and the EIS officers immediately met with local health officials to review medical records, laboratory data, and autopsy reports. The clinical picture was puzzling. The disease began with a mild prodrome (early symptoms that precede a serious illness, or syndrome) of fever, shortness of breath, achiness, and sometimes vomiting and abdominal pain, followed by a crisis phase in which the patient's lungs fill with fluid. This pattern was observed in the young woman who died at Crownpoint, her fiancé, and the girl who collapsed the night before. The girl's mother reported that the teenager had been under the weather for a few days, as if having a bad cold, but had felt well enough to join her friends on Friday night.

On Memorial Day, Butler visited the ICU at the University of New Mexico (UNM) hospital, which was admitting only persons with *This Mystery Disease*, or *TMI*, as the medical residents called it. They were extremely worried about contagion. Butler was struck by how very sick the patients were. He also noticed that those who survived the crisis recovered very quickly, unlike people who recover from septic shock due to bacterial infection. Those survivors included two close relatives of the woman who died at Crownpoint—her brother and pregnant sister-in-law[1, 2]—who had fallen ill less than two weeks after joining their grieving family in Little Water, where they stayed in the home in which their sister had lived with her fiancé.[2] The sister-in-law almost died, and her baby was born at the UNM hospital, two months early.[2]

Butler's initial job was to answer questions from the Arizona, Colorado, New Mexico, and Utah departments of health and from IHS physicians who sought his advice on keeping hospital patients and healthcare workers safe from the new disease. He also helped the EIS officers set up a disease surveillance system for identifying new cases. One EIS officer worked with UNM colleagues to develop a clinical description of the new disease,[3] and the other worked with health department officials to ship clinical specimens to the Special Pathogens lab (SPL) at the CDC. Back in Atlanta, wearing full-face positive pressure masks, SPL scientists led by Pierre Rollin, head of the SPL pathogenesis section, irradiated the blood samples so they would not be infectious and distributed them to other CDC labs to be tested for toxins or for antibodies to bacterial, viral, parasitic, and fungi diseases. Some non-irradiated samples were sent to SPL's highest containment laboratories (called Biosafety Level 4 or BSL-4) for injection into chicken eggs and tissue culture dishes in hopes that something would grow out.

Butler's duties also entailed representing the CDC at media events, an area in which he had little experience. Although his first press conference went well, he had a difficult moment when a reporter asked: When will you know? Butler replied that in 1976 it had taken six months to identify the bacterium that caused Legionnaires' disease—the respiratory illness described in Chapter 4. He heard the reporter say under his breath: I didn't know that! Butler also attended a late-night media event held by Navajo Nation President Petersen Zah and New Mexico Senator Pete Domenici in Window Rock, the seat of government of the Navajo Nation. Afterwards, Butler made his first visit to the IHS hospital in Crownpoint, where he met Hennessy, Golnick, and other local physicians.

The Crownpoint hospital was flooded with patients who feared they had the fatal mystery disease. The staff took a blood sample from each patient and arranged for a public health nurse to make follow-up calls. Unfortunately, the blood work did not help them distinguish between the "worried well"—people with colds—and people who were in danger of sudden respiratory failure. The only clue was that the patients with the mystery disease had a low platelet count (thrombocytopenia). When Hennessy suspected a patient might be in danger, he arranged for immediate transfer to the ICU at the UNM hospital. Sometimes he had to plead for a helicopter—I'm calling from Crownpoint, New Mexico, where three people have already died—and once he called an attending physician and went over the head of an emergency room doctor who refused to admit a patient. After that, he called the UNM physicians directly when he wanted a patient admitted.

By the end of the first week, some possible good news arrived from Atlanta: Special Pathogens Lab had a tentative diagnosis. After distributing irradiated specimens to other CDC laboratories, virologist Thomas Ksiazek had run antibody tests against a panel of SPL's own exotic "bugs," using reagents he and Peters and Rollin had developed at the Army Medical Research Institute of Infectious Diseases (USAMRIID) before they joined the CDC. The antibody tests were a longshot, because the SPL bugs were mostly hemorrhagic fever viruses not seen in the United States. Examples included the filoviruses that cause Ebola and Marburg hemorrhagic fevers, the arenavirus that causes Lassa fever, and the bunyavirus that causes Crimean-Congo fever. None of these pathogens cause the precise set of symptoms observed in the patients in Four Corners.

To everyone's surprise, Ksiazek's lab immediately reported hits on a rodent-borne bunyavirus called a hantavirus.[6] This did not make sense, because the only known disease caused by hantavirus, hemorrhagic fever with renal syndrome (HFRS), affects the kidneys, not the lungs. Moreover, HFRS—which was first identified in the 1950s among U.S. soldiers in Korea—has never been reported in the Western Hemisphere. Also, the antibody reactions were not strong. A weak signal might mean a false positive, but it could also mean a partial (but not identical) match to the hantaviruses on the SPL panel. Those included Hantaan and Seoul viruses (which cause HFRS), Puumala virus (which causes a milder form of kidney disease), and Prospect Hill virus (which is not associated with human illness). The virus that reacted most strongly to the antibody test was the Prospect Hill hantavirus, which had been isolated from a meadow vole in Frederick, Maryland.[7]

Nearly everyone at the CDC doubted Ksiazek's results, at least at first. Some thought it had to be a mistake, or a cross-reaction with a different virus. Others were skeptical, or alarmed, at the idea of a hemorrhagic fever virus—*like Ebola*!—in the United States. But for his part, Butler was relieved. His greatest fear was a 1918-like flu virus spreading like wildfire through the country and the world. Dangerous as they are, hantaviruses do not spread from person to person or cause pandemics. They are transmitted from rodents to people and do not spread beyond the habitats of the animals that carry them. This was no small thing. If the virus were really a hantavirus—and if it behaved like other known hantaviruses—Hennessy, Tempest, and their colleagues could be reassured that the disease would not spread within their medical facilities.

As it happened, a virologist who had studied the distribution of the Seoul hantavirus within the rat population of Baltimore[8]—Jamie Childs—was now working at the CDC in the division of viral diseases. Childs, an independent-minded and outspoken person who calls things as he sees them, was also skeptical, but for a different reason. He and his colleagues at Johns Hopkins and USAMRIID had spent years searching without success for evidence that rodent-borne hantaviruses cause acute human illness in the United States.[9] Nevertheless, Childs realized that if Ksiazek's results were real, the next step was to look for a rodent host. He left for New Mexico the next day.

Encouraged by this new development, the New Mexico Department of Health announced the SPL findings to the press, despite misgivings at the CDC about basing an announcement on antibody tests alone. No virus had grown out in petri dishes or animals in the BSL-4 laboratories. It fell to Stuart Nichol, a new member of SPL, to confirm the diagnosis, using the polymerase chain reaction (PCR), a powerful molecular technique that allows scientists to amplify and study vanishingly small amounts of genetic material. Invented in 1983 by Nobel Prize winner Kary Mullis, PCR is best known today for its forensic use in matching suspects' DNA to blood, hair, or sperm specimens found at the scene of a crime. But it is also a major diagnostic and research tool used to study the genetics of humans, animals, plants, and single-celled organisms. It can also be used to study viruses, which are typically made of RNA or DNA molecules covered by a protein coat, and to detect viral genetic material in a patient's blood or tissue.

As diagrammed below, the PCR technique can amplify any segment of DNA that lies between two DNA probes (or "primers") that are complementary in structure to the native DNA. Once the gap between the two primers is filled in by an enzyme called a polymerase, the process is repeated multiple times, creating more and more copies, faster and faster, as the chain reaction builds. Given the right primers, PCR makes it possible to detect and sequence any targeted stretch of DNA.

Early in his career, Nichol—an engaging and articulate Cambridge-educated virologist who grew up near Newcastle, in the North of

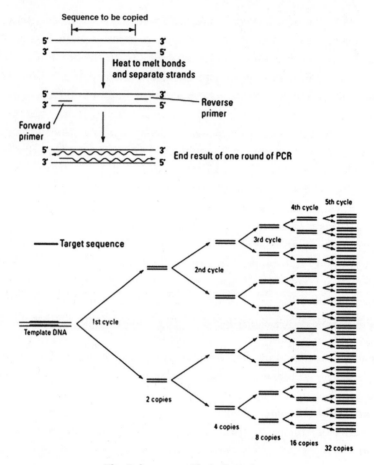

The Polymerase Chain Reaction

PCR targets the gene to be copied with primers, single-stranded DNA sequences that are complementary to sequences next to the gene to be copied. To begin PCR, the DNA sample that contains the gene to be copied is combined with thousands of copies of primers that frame the gene on both sides. The DNA polymerase enzyme uses the primers to begin DNA replication and copy the gene. The basic steps of PCR are repeated over and over, creating billions of copies of the DNA sequence between the two primers.

England—recognized the power of the new technique to study viral replication and mutation under natural (i.e., non-laboratory) conditions.

As a professor at the University of Nevada, he had used PCR to follow the near real-time evolution of a cattle pathogen called vesicular stomatitis virus over the course of a naturally-occurring outbreak.[10] To work with vesicular stomatitis virus, which is an RNA virus, Nichol had employed a PCR technique called reverse-transcription PCR (RT-PCR) that translates RNA into DNA before the amplification steps begin. Now Peters asked him to use RT-PCR on clinical samples from Four Corners. Peters bet him that the Four Corners pathogen would turn out to be the Prospect Hill hantavirus.

Nichol was well aware of the atmosphere of urgency and frustration at the CDC. That night, he sat up until 4 AM to design PCR primers that would match up with sequences from known hantaviruses. His parents were visiting from England, and after they went off to bed he sat down with a print-out that aligned coat-protein gene sequences from the Hantaan, Seoul, Puumala, and Prospect Hill viruses. He looked for conserved sequences common to two or more viruses that bracketed a small region of greater divergence (the part that would be amplified). He worked on sheet after sheet, looking for appropriate sequences. When he was done, he had designed two sets of primers, one based on gene sequences from the Hantaan and Seoul hantaviruses and one based on sequences from the Prospect Hill and Puumala viruses. However, his lab's first attempt to amplify a hantaviral sequence did not succeed. The problem could be a lack of viral RNA in the patient sample, or viral RNA that was in poor condition. Or, it might be that the PCR primers were not closely enough matched to the viral sequences. To overcome these obstacles, Nichol added an additional PCR step involving "nested" primers. He performed a first round of PCR using primers that surrounded a large stretch of DNA and then a second round using internal primers to amplify a smaller internal fragment. This was like enlarging a picture with a Xerox machine and then enlarging a portion of the Xeroxed picture over again.

Voila! After a second round of amplification using a nested set of Prospect Hill/Puumala primers, there it was—-a very faint band of DNA of the expected size. Was this real? The fragment was carefully excised, sequenced, and analyzed. It was definitely a hantavirus, but not an exact match to Prospect Hill.[11] Could it be an entirely new virus—a novel, disease-causing hantavirus carried by rodents in the United States?

Seeking confirmation, Hughes and Peters described the antibody and PCR findings over the phone to hemorrhagic fever experts at USAMRIID and Johns Hopkins University, who agreed that the data pointed to a hantavirus. Soon after, CDC pathologist Sherif Zaki provided convincing visual evidence by probing clinical specimens with anti-hantaviral antibodies generated in mice. Using a fluorescent stain, he demonstrated that the patients' lung tissues were full of hantaviral proteins, especially along the walls of tiny blood vessels (the capillaries).[12, 13] The viral proteins apparently made the lung capillaries more permeable, causing them to leak plasma. In this way, the pathogenesis of the illness differed from that of other hemorrhagic fevers, which typically involve leakage of both cells and plasma (i.e., hemorrhage) in many parts of the body. Convinced, the CDC scientists named the illness hantavirus pulmonary syndrome, or HPS.

What did this discovery mean for the people in Four Corners? To protect themselves and their families they would need to know how and why the virus was spreading. Was there a popular pastime that was bringing people into greater contact with mice? Had a new rodent species entered the area, like the deer that brought Lyme disease to suburban New England as their forest habitats grew smaller? Pondering these questions, Butler headed for the town of Gallup, near the center of the affected area, to lead an investigation of factors that put people in danger of acquiring the deadly illness. He was joined by Jamie Childs, the CDC expert in rodent hantaviruses, who assembled a team to investigate the rodent population.

Known as the "heart of Indian Country," Gallup is a mecca for tourists and movie-makers, surrounded by semi-arid Western landscapes of mesas, canyons, and wide open spaces. Butler checked into a local hotel that was hosting the cast and crew of Oliver Stone's *Natural Born Killers*. The next day he had diarrhea and mild fever that he blamed on eating an enchilada. Though he knew it was probably a stomach virus, he could not help worrying. He was a pretty good runner, and he knew that some runners had died, people like him who were young, fit, and active. He kept reassuring himself that *vomiting is not part of the clinical picture* and *my breathing is fine!* Fortunately, after a good night's sleep he felt much better.

Over the next few days, more than twenty public health colleagues joined Butler and Childs in Gallup. Among them was Joseph McDade—the revered scientist who solved the Legionnaires' Disease mystery described in Chapter 4—who came to New Mexico at Hughes' request to meet with President Zah and enlist his personal support and help. Hughes also recruited Patrick McConnon, a veteran of many other public health crises (as described in Chapter 2), to advance the field investigation by ensuring that CDC staff (who were working out of a trailer) had lodgings, work space, office equipment, and field supplies. McConnon rented an office near the IHS hospital and arranged for computers, telephones, fax machines, a photocopier, and dry ice. His job was not easy because Gallup, though rugged and beautiful, was also, in economic terms, poor and underdeveloped. He was grateful for the help of Navajo Nation and IHS colleagues such as Bernice Melani-Baca, a resourceful Zuni woman who knew everyone in town and could find anything her colleagues needed.

Officials from the Navajo Nation and the IHS arranged an orientation on Native American attitudes, religious sites, and historic events, including the 1864 Long Walk that exiled the survivors of the Navajo Wars from their ancestral lands. Butler remembers it as a "crash course" in Native American culture; Childs remembers it as what today we call "sensitivity training." McConnon thought it was a remarkable learning opportunity for the CDC investigators: an introduction to a complex universe of cultural and spiritual knowledge that outsiders could never fully grasp. While he hoped the local community would forgive their unintended offenses, it was valuable to be aware of obvious (and avoidable) pitfalls. According to Navajo tradition, for example, the names of the dead should not be written or spoken for four days after death, to give the spirit time to pass from the earthy plane to the spiritual one. Also, the Navajos believe that blood taken for diagnostic purposes remains spiritually attached to the body and its unauthorized use by scientists is therefore a violation of privacy and autonomy.

The Navajos were also unhappy at being linked to the disease in Four Corners. The newspapers not only printed the names of dead relatives, but also referred to the disease as the "Navajo flu," stigmatizing an entire people. Navajos families were turned away from restaurants,[4, 14] and Navajo children on a trip to California to visit pen-pals were barred from entering a school.[15]

Butler and Childs worried that the Navajo tribal authority would not allow them to conduct an investigation on Navajo land. The community was traumatized by the deaths of their young people and angry at the flood of journalists and officials who violated their customs and reinforced their long-standing distrust of the U.S. government. Many suspected that the new disease was a result of germ warfare—either an accidental release of a dangerous microbe or an experiment in which Navajos were used as guinea pigs. During their deliberations, Navajo officials invited Butler and Childs to a council meeting at which a tribal elder recited a litany of atrocities: "You sent us on the Long Walk, you put us on a reservation, you gave our people smallpox. . . . Why should I think you are not the one who also brought this disease to us?" Butler recalls this speech as a statement of distress that vividly expressed the feelings of many people in the local community.

In the end, however, the tribal authorities apparently decided that the benefits of solving the mystery outweighed their misgivings. President Zah stipulated that helicopters and other large vehicles be kept off Navajo land, but allowed staff from the Navajo Nation Division of Health to participate in the investigation. He also asked the press to stay away.

Based on what they knew about hemorrhagic fever with renal syndrome, the medical detectives had begun generating hypotheses about activities that could facilitate disease spread. Anxious to test them, Butler's team worked all night to develop a standard questionnaire for interviewing surviving HPS patients and the family members of those who had died. Their study population included seventeen people, including four survivors, who had fallen ill between mid-March and mid-July.[16] (The two surviving patients in the family of the Little Water couple decided not to participate.) Twelve of the seventeen were Native

Americans, four were white, and one was Hispanic. Sixteen had lived in rural areas, and one lived in a small town but had visited a rural area each weekend. For comparison, they interviewed healthy people from three control groups: the patients' own households (household controls), neighboring households (near controls), and households located some distance away (far controls). These households were matched by location (rural or urban, on or off the reservation).

Here is what they found: More patients than controls had trapped rodents, handled dead mice, or observed rodents or rodent droppings near their homes. One patient who died had been scratched by a mouse. The patients were also more likely to hand-plow their fields using a shovel or a hoe, to plant crops, or to report their occupation as "herder." Finally, the patients were more likely than their family members to clean outbuildings where food, grain, or hay was stored, or where animals were kept. No risk was associated with hunting, fishing, eating home-butchered meat, participating in traditional ceremonies, or having insect bites or contact with dogs or cats. Another hypothesis that did not pan out was that people had become infected while digging up piñon nuts hoarded by woodrats.

What did trapping mice, hand-plowing, and shed-cleaning have in common? All of them can increase the risk of lung infection through inhalation of contaminated dust particles containing dried-out rodent droppings, urine, or saliva. People who trap rodents, for example, can stir up infected particles when placing traps in dusty corners with poor ventilation or when sweeping out sheds, which can propel infected particles into the air. Moreover, herders may be at special risk during their early-spring custom of re-opening "sheep camp" hogans that are closed during the winter. Handling or being scratched by a mouse might also cause infection through inhalation, or through touching the mucous membranes of one's nose or mouth with a contaminated finger.

While Butler's team completed the interviews, Childs' team conducted a formal inspection of homes and family compounds.[19] They found the homes of the dead deserted, with unmade beds and dirty dishes as visible signs of lives cut short. Otherwise, the investigators found few differences between the patients' homes and those of the near and far controls. Members of both groups had lived in wood-frame or adobe houses, hogans, or trailers whose yards and outbuildings contained materials that rodents could use for nesting (discarded tires, car bodies, or wood piles) or food (cereal, pet food, or grain stored in sacks).

When given permission, Childs' team placed traps inside and outside people's houses, baited with oats and crushed corn. They used small traps for catching mouse-sized rodents and medium ones for squirrel- or rat-sized rodents. A Navajo environmental health expert named Herman Shorty eased their way in working with the local people, many of whom were interested and supportive. Shorty was a cheerful, stocky, good-humored person who was devoted to Navajo traditions. Everywhere they went, he knew whose land they were on, and whether it was all right to be there. He also knew the terrain and how to get the CDC trappers where they needed to be.

McConnon called the trappers "cowboys" because of their colorful and independent ways. They included Gary Maupin and Kenneth Gage, from the CDC laboratory in Ft. Collins, Colorado, who knew the local fauna very well because of their ongoing field work on plague. Maupin and Gage rolled into Gallup in an ancient, beat-up four-wheel drive Ford Bronco that somehow lasted through the investigation and the trip back to Ft. Collins. They liked to play tricks, tell tall tales, and have a good time. As Childs says, people who hunt dangerous animal viruses for a living tend to be crazy guys who work hard and play hard. (The stereotypical epidemiologist, in contrast, tends to be quiet, sensible, and diplomatic.) Within a few intense days the cowboys had formed a special bond. They worked long hours, without a break, getting up at 4 AM each morning to retrieve the traps set the night before and load them into their battered old Bronco.

Like people who work in BSL-4 laboratories, the trappers had to be extremely focused and careful when working in the field, even if they let loose after hours. They stood upwind when opening traps and wore respirators, gloves, and full-body suits, despite the intense summer heat. Once they had completed their trap-collection rounds, they would set up camp to process the animals and record their species. They harvested body parts from each rodent—an eye, the spleen, the lungs, and the liver—and packaged them on dry ice for shipment to Atlanta. Afterwards they returned the rodents to nature by burying them.

The cowboys disliked being followed by reporters, who were around all the time, seeking scoops and stories, in complete disregard of President Zah's request. Processing the animals was labor-intensive and dangerous, and they did not want to be disturbed or photographed in their weird-looking protective suits. Once they scared off a reporter by telling him he was downwind of infected rodents. Another time they sent a decoy truck in the wrong direction, followed like a pied piper by a line of cars filled with photographers and journalists.

The cowboys' work had three major results. First and foremost, they documented that the homes of the HPS patients had significantly more rodents than the homes of healthy controls.[16, 17] That explained why the HPS patients were more likely to trap mice—the homes of the Little Water couple and the other HPS patients were densely infested. Mice had burrowed into the walls of hogans (which were made of logs and packed with mud and earth), as well as into the insulation panels of trailers. However, antibody testing by SPL indicated that individual rodents in the patients' homes were no more likely to be infected than rodents in the control homes. The percentage of infected rodents in both cases was about one-third. Thus, the danger lay in having large numbers of rodents in one's home. The second big result, made in collaboration with Nichol's laboratory, was that viral gene sequences amplified from the blood or tissue of an HPS patient were virtually identical to sequences amplified from mice trapped inside that patient's home.[11] This finding confirmed beyond a reasonable doubt that the new hantavirus was the cause of the outbreak. It was also a direct demonstration that DNA "fingerprints" made by PCR can match up individual patients and pathogens.

The third result caused some dismay. The natural reservoir of the new virus turned out to be the most abundant small mammal in most of North America: the deer mouse *Peromyscus maniculatus*.[17, 18] These cute little animals—only three to four inches long, not including the tail—are fearless creatures who thrive in human dwellings, sheds, and barns. They are everywhere, and cannot be eliminated or easily avoided. Another *Peromyscus* species that tested positive was the white-footed deer mouse (*Peromyscus leucopus*), which is common in the northeastern United States and also carries the ticks that cause Lyme disease. Other rodent species that sometimes carried the hantavirus included the house mouse, piñon mouse, and brush mouse; the cliff and Colorado chipmunks; the desert cottontail; and the white-throated woodrat.

Based on these findings, the CDC issued public health recommendations that emphasized avoiding mice and preventing them from entering your home.[19] The people of Four Corners were advised to remove potential food sources and nesting sites and to use special precautions, like wearing a mask or moistening a dusty floor with water before cleaning out sheds or other areas infested with rodents. Campers and hikers were advised to keep food in rodent-proof containers and not to pitch their tents in areas near rodent burrows or woodpiles that could serve as rodent shelters.

While some Navajos questioned the idea that a common everyday mouse could carry disease, others began adopting cats as pets, despite a marked cultural preference for dogs. It helped that the CDC's recommendations echoed those of local medicine men and tribal healers who urged their communities to destroy belongings contaminated by mice droppings. The medicine men said that clusters of respiratory disease had occurred at least twice before in the twentieth century, in years when heavy rains led to large piñon nut harvests and huge numbers of deer mice.[20] Ecologic data published by UNM mammologist Robert Parmenter backed-up this scenario for 1993, documenting that luxurious plant growth following heavy snow and rainfall had led to an over-abundance of piñon nuts and grasshoppers, which in turn stimulated a ten-fold rise in the local rodent population.[21] Parmenter traced this chain of ecologic events—called a "trophic cascade"—back one step further to a 1991–1992 El Nino event that apparently caused the wet winter and abundant spring rain.[21, 22]

By the middle of August the outbreak was essentially over, although occasional cases continued to be reported, typically when someone swept out a storage area without taking precautions. Twenty-four residents of Four Corners had fallen ill, and half of them had died. Butler spent two more weeks traveling in the Southwest to spread the word about disease prevention and share what he knew about patient care. Back in June, he and Peters had considered initiating a treatment trial of ribiviran, a not-yet-licensed antiviral medicine used to treat Lassa fever[23] and HFRS.[24] However, Hennessy warned them that conducting a randomized trial would be seen as experimenting on the Navajo people. Instead, they arranged for ribavirin to be provided on request to any person with suspected HPS.

Although ribavirin proved ineffective against HPS,[25] the ICU staff at UNM learned how to help patients survive the crisis phase through improved supportive care. They knew, from painful experience, that providing fluids by IV made things worse, causing more fluid accumulation in the lungs. Instead, they performed a procedure called "extra corporeal membrane oxygenation" in which a person's blood is oxygenated outside of the body and then transfused back.[26] This ICU procedure is still used to save the lives of people with HPS.[27]

Looking back, Hennessy remembers the terrible day in May when the twenty-one-year-old woman died as one of the worst in his life. He left New Mexico the following year to join the EIS program, as described in Chapter 6. Two of his friends still work at the Crownpoint hospital, where they continue to report a few HPS cases each year. They are equipped with modern diagnostic assays, and they transport anyone who tests positive to UNM, which remains a national leader in HPS treatment. Moreover, they continue to educate their patients about avoiding exposure to mouse droppings, especially in years with above-normal spring rains. When it comes to HPS, prevention is the most important thing.

Butler also speaks with deep feeling of his experience in Four Corners. It was in his mind, ten years later, when another infectious threat, severe acute respiratory syndrome (SARS), spread from Asia to Europe and North America, with several confirmed cases in the U.S. and sizable outbreaks in Canada, Taiwan, and Singapore.[28] Hantavirus pulmonary syndrome might as easily have been called severe acute respiratory syndrome. In both cases, no one knew what type of pathogen was involved, so public health authorities tested for everything. And in both cases, there was intense concern about contagion, especially in healthcare facilities. Although SARS had a lower fatality rate than HPS, it was more difficult to control, because it was transmissible from person-to-person, though without the ease and rapidity of pandemic influenza.

Unlike SARS, HPS is not really a new disease—just a newly identified one. As suggested by the medicine men's historical account, HPS has probably caused occasional unexplained deaths in North and South America—throughout the deer mouse habitat—for hundreds of years. Once HPS had been discovered, sporadic individual cases caused by the HPS virus or closely related viruses (including some from before 1993[29]) were reported in many parts of the country.[30] It was only because a few cases of sudden respiratory failure were reported by astute local physicians that the disease was investigated and characterized. If not for their expertise and vigilance, HPS might have gone undetected much longer. As it happens, HPS clusters are very rare. In fact, no other cluster was reported until 2012, when ten cases (three of them fatal) occurred among tourists at Yosemite National Park who stayed in cabins infested with deer mice, following another wet winter and lush spring.[31, 32] It was also fortunate that by 1993, the CDC had begun upgrading its diagnostic capacities following a decade of decreased resources.[33] (Ironically, the new resources were justified as needed to protect the public from the introduction of "foreign" diseases like Ebola and AIDS—not from a new disease detected in the United States!) Without Ksiazek's hemorrhagic fever reagents and Nichol's knowledge of RT-PCR, the HPS investigation would have stalled (like the Legionnaires investigation in 1976) until a way was found to grow the new virus in the laboratory.

The CDC and USAMRIID announced the isolation of the HPS hantavirus in November 1993, six months after the first cases were identified.[34] When an animal-borne pathogen is isolated for the first time, the custom is to name it after a geographic feature in or near the place where it was found. Ebola was named after the Ebola River in the Republic of the Congo, Puumala after a town in Finland, and

Prospect Hill after an estate in Maryland. In this case, however, the Navajo community was uncomfortable with names that would tie them to the new disease, such as the *Four Corners virus,"* or the *San Juan virus* after a river whose left bank lies within tribal lands. As an alternative, the SPL scientists proposed *Muerto Canyon* (Canyon of Death), after a small arroyo, thinking it would be both appropriate and inoffensive. But it turned out that the Spanish form of this name has special historical meaning for the Navajos: the Canyon del Muerto is a branch of the Canyon de Chelly, the site of the last major battle of the Navajo Wars, which ended with the mass destruction of Navajo settlements and the start of the Long Walk.

In the end, the scientists settled on a name that, though chosen to be neutral and innocuous, managed to convey the mystery and menace of an unknown threat. The name of the pathogen that causes HPS is *Sin Nombre*, the No-Name Virus.

Epilogue

June, 2013

Because new microbes continue to emerge, we must always be prepared for the unexpected.

As this book goes into production, international investigations are underway—one in the Middle East and one in China—involving new respiratory diseases with high fatality rates and the potential for global spread.

Thus far, we know that the outbreak in the Middle East is caused by a previously unknown virus that is genetically related to the coronavirus that caused the global outbreak of severe acute respiratory disease (SARS) in 2003. As of June 2013, the new virus—called the **Middle East respiratory syndrome coronavirus (MERS-CoV)**—has been laboratory-confirmed as the cause of fifty cases and thirty deaths, most of them in Saudi Arabia, Qatar, Jordan, and the United Arab Emirates. Three cases were reported in the United Kingdom, in a person who had traveled to Saudi Arabia and two of his family members, as well as two in France, in a person who had traveled to the United Arab Emirates and a patient who shared the sick man's hospital room. Three more were reported in Tunisia, including a fatal case in a man who had visited Qatar and Saudi Arabia, and mild cases in two of his adult children. New cases continue to be reported in Saudi Arabia, some of them related to an ongoing outbreak in a healthcare facility. Although the source of MERS-CoV is as yet unknown, scientists suspect that it is carried by animals. The closely-related SARS virus, whose host animal is the Chinese horseshoe bat, most likely infected humans in 2003 via infected civets (catlike mammals) sold for food in live animal markets.

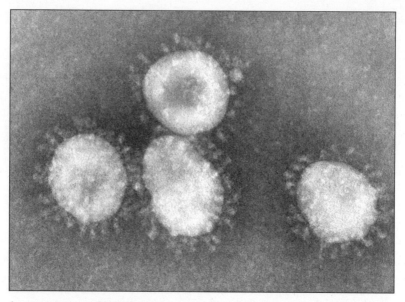

A coronavirus is an RNA virus with a protein envelope. Its name is derived from the Latin *corona*, meaning crown or halo, and refers to bulbous projections on the surface of the virus. *Source:* Image courtesy of the CDC's Public Heath Image Library

The outbreak in China involves a new strain of influenza in poultry—**avian influenza A(H7N9)**—that has caused 132 cases of human disease and 31 deaths in eight Chinese provinces, Beijing, Shanghai, and Taiwan. For reasons as yet unknown, most cases have occurred in men, with few children falling ill. The suspected source of infection is infected birds such as chickens and ducks. However, the H7N9 virus typically does not cause symptoms in birds, which makes its presence in poultry and wildlife difficult to track.

Fortunately, there is no evidence that either the novel coronavirus or the new strain of H7N9 influenza is capable of sustained person-to-person transmission. (Thus far, MERS-CoV has spread only to close contacts, and H7N9 influenza has spread from infected poultry or contaminated environments.) Nevertheless, public health authorities are playing close attention, fearing that H7N9 or MERS-CoV could evolve into an easily transmissible virus that spreads rapidly around the world. World Health Organization (WHO) member countries are required to report new cases of human infection with these viruses to WHO as part of their obligations under the 2005 International Health Regulations. As isolates of these viruses are obtained from new patients, public health scientists (including virologists from the U.S. Centers for Disease Control and Prevention (CDC)) are examining them for genetic changes that could lead to greater virulence or greater transmissibility.

Notable outbreak investigations are also underway in the United States, including one involving an unusual fungal infection that has sickened more than 730 people in twenty states, causing 52 deaths. The first cases were detected more than a year ago, in Tennessee, and new cases—about ten a week—continue to be reported. The source of the outbreak was traced to three lots of a medical steroid product made by a compounding pharmacy in Massachusetts that was injected into patients' spines or joints to treat lower back pain, joint pain, or sciatica. The contaminant turned out to be *Exerholium rostratrum*, a common mold found in soil and on plants that rarely cause human disease. However, fungal infection acquired by direct injection—a most unusual route—has caused meningitis, spinal and joint infections, and strokes. CDC has convened a group of scientific experts to provide treatment advice to local physicians, most of whom have never seen a case of infection with *Exerholium rostratratum*, and the Food and Drug Administration is reviewing the need for better regulation of compounding pharmacies, which typically customize medical products to fit the needs of individual patients rather than producing them in bulk, as in this case.

Across the nation, medical detectives continue to investigate whatever new or re-emerging threats arise. These range from newly discovered tickborne pathogens in Missouri (the Heartland phlebovirus) and in Wisconsin and Minnesota (a new species of *Ehrlichia* bacteria) to dengue fever in the Florida Keys, newly emerged after a hiatus of sixty years. Other diseases under investigation in June 2013 include West Nile fever in Dallas County, Texas (the epicenter of a multi-state outbreak in 2012), salmonellosis in Fayetteville, North Carolina (traced to food served at a hotel), Rocky Mountain Spotted Fever on tribal lands in Arizona (carried by an unexpected species of dog tick), Valley Fever in California's Central Valley, and drug-resistant tuberculosis in downtown Los Angeles. U.S. medical detectives are also working with colleagues at the Pan American Health Organization (PAHO) to keep the Americas free from measles, by detecting imported cases and tracing their contacts. Efforts are also underway to ensure public health readiness to detect and control exotic mosquito-borne diseases like Chikungunya fever (pronounced chi-koon-goo-nya) that—like West Nile disease in 1999—could be introduced into the Western Hemisphere.[1]

Medical detectives in nearly all states and localities are also concerned about outbreaks of hard-to-treat bacterial infections that patients acquire while receiving medical treatment for other conditions. These infections are called healthcare-associated infections, or HAIs. Although significant progress has been made in reducing the incidence of HAIs, using innovative infection control protocols and checklists, outbreaks continue to occur in hospitals and other healthcare settings. HAIs of special concern include those caused by *Clostridium difficile* and drug-resistant bacteria like methicillin-resistant Staphylococcus aureus (MRSA) and carbapenem-resistant Enterobacteriaceae (CRE), which can cause life-threatening complications. Medical detectives are applying new epidemiologic and laboratory tools that can help detect and control individual cases and small clusters of HAIs before they develop into large outbreaks. This is difficult work, requiring both ingenuity and persistence, because of the diversity of the bacterial species that cause HAIs and the intensive efforts needed to identify early cases and institute effective control measures.

Because the battle between humans and microbes is unpredictable and never-ending, the outbreak stories of the future are likely to involve new pathogens and new scientific puzzles. New diseases will emerge from animal reservoirs, while known ones will resurge, sometimes in drug-resistant forms. I am confident (and grateful) that the next generation of medical detectives will continue to protect the nation's health by meeting new challenges with intelligence, focus, and determination.

Where Are They Now?

James W. Buehler, MD: After completing his training at the Epidemic Intelligence Service, as described in Chapter 4, "Sorrow and Statistics," Dr. Buehler joined the CDC as a commissioned officer in the U.S. Public Health Service, where he focused on maternal and child health, HIV/AIDS, and population health monitoring. In 2010, he became the founding director of the CDC Public Health Surveillance Program Office, which is now called the Public Health Surveillance and Informatics Program Office. After retiring from CDC in 2013, Dr. Buehler became a Professor of Health Management and Policy at the Drexel School of Public Health.

Jay Butler, MD: After investigating the outbreak described in Chapter 7, "A Normal Spring," Dr. Butler continued his work on respiratory infections at the CDC National Center for Infectious Diseases in Atlanta until 1998 and in Anchorage, Alaska, until 2005. In 2005 he was appointed Alaska State Epidemiologist, and in 2007 he became Chief Medical Officer for the State of Alaska. Since 2010, he has been a member of the executive team of the Alaska Native Tribal Health Consortium in Anchorage, where he oversees statewide public health services in the tribal health system, serves as Alaska Native Medical Center's Medical Director for Infection Control, and works at the state and national level to support tribal self-determination for improving health and wellness throughout Indian Country.

Jamie Childs, PhD: After leading the rodent field investigation described in Chapter 7, "A Normal Spring," Dr. Childs continued to work at the CDC as chief of the Viral and Rickettial Zoonoses Branch at the CDC and at the Johns Hopkins School of Public Health as an associate professor. In 2004, Dr. Childs joined the Yale University School of Public Health, where he studies the ecological dynamics of zoonotic viruses and vector-borne bacteria.

Duane J. Gubler, ScD. MS: Dr. Gubler, who helped solve the medical mystery in Chapter 1, "Dead Crows Falling from the Sky," is an internationally recognized expert on dengue fever. Dr. Gubler was the founding Chief of the CDC Dengue Branch in Puerto Rico and served as Director of the CDC Division of Vector-Borne Infectious Diseases in Fort Collins, Colorado, for fifteen years before moving to Honolulu to chair the Department of Tropical Medicine, Medical Microbiology and Pharmacology at the John A. Burns School of Medicine of the University of Hawaii. In 2007 he became the Founding Director of the Signature Research Program in Emerging Infectious Diseases at the Duke-NUS Graduate Medical School in Singapore.

Craig W. Hedberg, PhD: Dr. Hedberg, whose activities as a field epidemiologist are described in Chapter 5, "Dangerous Desserts," continued to investigate medical mysteries in Minnesota for another five years, while also serving as supervisor for foodborne disease surveillance at the Minnesota Department of Health. Today, he is a professor in the Division of Environmental Health Sciences at the University of Minnesota School of Public Health and an internationally recognized expert on public health surveillance and the control of foodborne diseases.

Thomas Hennessy, MD, MPH: After working as a family physician on the Navajo Indian Reservation, as described in Chapter 7, "A Normal Spring," Dr. Hennessy joined the CDC Epidemiologic Intelligence Service and was assigned to the Minnesota Department of Health, where he participated in the field investigation described in Chapter 5, "Dangerous Desserts." Today, Dr. Hennessy is the Director of the CDC Arctic Investigations Program, which promotes the health of the people of the Arctic and sub-Arctic through disease surveillance, epidemiologic research, and development and evaluation of vaccines and other public health interventions.

James M. Hughes, MD: As Director of the National Center for Infectious Diseases from 1992 to 2005, Dr. Hughes oversaw the CDC's response to many complex public health emergencies at home and abroad, including the outbreak described in Chapter 7, "A Normal Spring." Dr. Hughes currently serves as Professor of Medicine and Public Health at the School of Medicine and the Rollins School of Public Health at Emory University.

Ruth Lynfield, MD: Since solving the medical mystery described in Chapter 6, "The Red Mist," Dr. Lynfield has continued to serve as Minnesota State Epidemiologist and Medical Director for Infectious Diseases. She also holds adjunct faculty appointments in Internal Medicine, Epidemiology, and Pediatrics at the University of Minnesota.

James S. Marks, MD, MPH: After completing his training at the Epidemiologic Intelligence Service, as described in Chapter 5, "Obsession or Inspiration," Dr. Marks joined the CDC in Atlanta, where he identified systematic ways to prevent and detect cancer, heart disease, and diabetes, and to reduce obesity and tobacco use. He served for almost a decade as director of the National Center for Chronic Disease Prevention and Health Promotion before joining the Robert Wood Johnson Foundation as Senior Vice President and Director of the Health Group. In this capacity Dr. Marks oversees the Foundation's work on childhood obesity, public health, and vulnerable populations.

Patrick McConnon, MPH: Over the course of more than thirty-five years at the CDC, Mr. McConnon—whose activities are chronicled in Chapter 2, "The McConnon Strain," and Chapter 7, "A Normal Spring"—undertook challenging assignments in many areas, including refugee health, smallpox and Guinea worm eradication, and domestic and international outbreak coordination. From 2002 to 2012 he served as the Executive Director for the Council of State and Territorial Epidemiologists.

Michael T. Osterholm, PhD, MPH: After solving the medical mystery described in Chapter 5, "Dangerous Desserts," Dr. Osterholm continued to serve as the Minnesota State Epidemiologist for seven more years. Today, he is an internationally recognized expert in a number of areas of infectious disease epidemiology, including emerging infections, food safety, and biosecurity. He is the Director of the Center for Infectious Disease Research and Policy (CIDRAP) at the University of Minnesota. Dr. Osterholm is also Distinguished University Teaching Professor in the University of

Minnesota School of Public Health, Professor, Masters of Science in Security Technologies, Technological Leadership Institute, University of Minnesota College of Science and Engineering and an adjunct professor in the University of Minnesota Medical School.

Steven Solomon, MD: After completing his training in the Epidemic Intelligence Service, as described in Chapter 4, "Sorrow and Statistics," Dr. Solomon joined the CDC in Atlanta, where he has focused on infection control, healthcare epidemiology, antimicrobial resistance, and health-systems research. He currently serves as Director of the Office of Antimicrobial Resistance in the Division of Healthcare Quality Promotion, National Center for Emerging and Zoonotic Infectious Diseases.

Richard N. Danila, PhD, MPH: Dr. Danila—whose work as a medical detective at the Minnesota State Department of Health is described in Chapter 5, "Dangerous Desserts" and Chapter 6, "The Red Mist"— currently serves as Minnesota Deputy State Epidemiologist and section manager of the Acute Disease Investigation and Control Section, with a special focus on public health planning for emergency preparedness and response. Dr. Danila is also a member of the adjunct faculty of the University of Minnesota School of Public Health, Epidemiology Division.

Aaron DeVries, MD: Dr. DeVries, a leader of the investigation described in Chapter 6. *The Red Mist,* is a medical epidemiologist in the Infectious Disease Epidemiology, Prevention, and Control Division at the Minnesota Department of Health. He is also an Adjunct Assistant Professor at the University of Minnesota in the Department of Medicine and the School of Public Health.

Annie Fine, MD: Dr. Fine, whose activities as a medical detective are described in Chapter 1, "Dead Crows Falling From the Sky," has continued to work at the New York City Department of Health and Mental Hygiene, where she currently serves as Medical Director, Data Analysis, Reportable Disease Surveillance and Bureau of Communicable Disease.

Stacy Holzbauer, DVM, MPH: Dr. Holzbauer, the veterinary epidemiologist whose work is described in Chapter 6, "The Red Mist," currently serves as a CDC Career Epidemiology Field Officer assigned to the Minnesota Department of Health.

Marcelle Layton, MD: Dr. Layton, who solved the medical mystery described in Chapter 1, "Dead Crows falling from the Sky," has played key roles during many other emergencies, including the public health response to 9/11 and to the anthrax incidents in New York City. She currently serves as Assistant Commissioner, Communicable Disease, at the New York City Department of Health and Mental Hygiene.

Joseph McDade, PhD: Twelve years after solving the medical mystery described in Chapter 4, "Obsession or Inspiration," Dr. McDade discovered a new species of *Ehrlichia* bacteria that causes a tickborne disease in the southeast and south-central United States. Dr. McDade retired from the CDC in 2003 and is currently advising pre-medical students, consulting with the CDC, and serving as an advisor to the *Emerging Infectious Diseases* journal, of which he was a founding editor.

Stuart T. Nichol, PhD: Dr. Nichol, whose work as a molecular virologist is described in Chapter 7, "A Normal Spring," has continued to work at the CDC, where his current position is Chief of the Viral Special Pathogens Branch at the National Center for Emerging and Zoonotic and Infectious Diseases.

James J. Sejvar, MD: Dr. Sejvar, the neuroepidemiologist whose work is described in Chapter 6, "The Red Mist," is a staff epidemiologist at the CDC National Center for Emerging and Zoonotic Infectious Diseases and an assistant professor of Neurology at the Emory University School of Medicine.

Chapter Notes

Introduction

1. Dr. Lederberg's saying is quoted in: Shalala DE. Collaboration in the fight against infectious diseases. Emerg Infect Dis. 1998 Jul-Sep;4(3):354-7. His ideas about "Our Wits Versus Their Genes" may be found in: Lederberg J. Infectious history. Science. 2000 Apr 14;288(5464):287-93.

2. This saying is variously attributed to Irving J. Selikoff (1915–92), who pioneered occupational medicine, and Sir Austin Bradford Hill (1897–1991), who pioneered the use of random clinical trials.

Chapter 1. Dead Crows Falling from the Sky

Chapter 1 draws on interviews with Marci Layton, Annie Fine, and Duane Gubler.

1. In most years, the NYS Arthropod-Borne Disease Program performs viral isolation studies by culturing material from statewide (or multi-county) mosquito surveillance pools in Vero cells, a line of African green monkey cells that supports the growth of many mosquito-borne viruses, including SLE virus. However, due to budget constraints, analysis of mosquito surveillance pools in 1998 and 1999 was limited to molecular (RT-PCR) testing for two encephalitis viruses: eastern equine encephalitis

virus and California encephalitis virus (John Howard, NYS health department, personal communication). The last time SLE had been detected in New York State was in 1975, a year in which SLE was also detected in several other states.

2. See: Kennedy R., "Man Versus Mosquito." *New York Times.* September 17, 2000.

3. In recent years, "syndromic surveillance" has acquired a second, specialized meaning in the public health community, referring to the use of electronic data to monitor increases in syndromes of interest.

4. United States General Accounting Office. Report to Congressional Requesters. *West Nile Virus Outbreak. Lessons for Public Health Preparedness.* GAO/HEHS-00-180, September 11, 2000. Page 41. Appendix II of this report provides a detailed chronology of the animal and human outbreaks. A timeline of the outbreak is also provided in *West Nile Fever: A Medical Detective Story.* Bio Bulletin of the American Museum of Natural History (http://www.amnh.org/education/resources/rfl/web/bulletins/bio/biobulletin/story1378.htm)

5. Jacobs A. "Exotic Virus is Identified in 3 Deaths." *New York Times.* September 26, 1999.

6. The minimum WNV infection rate in *Culex* mosquitoes in Brooklyn in fall 1999, may have been as high as 57 per 1000. See: Nasci RS, White DJ, Sterling H, Oliver JA, Daniels TJ, Falco RC, Campbell S, Crans WJ, Savage HM, Lanciotti RS, Moore CG, Godsev MS, Gottfried KL, Mitchell CJ. "West Nile virus isolates from mosquitos in New York and New Jersey, 1999." *Emerg Infect Dis,* 2001 Jul–Aug; 7(4): 626–30.

7. Hauer accused anti-spraying protesters of "irresponsible environmental hysteria and stupidity" [*Newsday,* 10/9/99, "Bugged by Spraying," by Dan Morrison], and Giuliani said that the protestors were "in the business of wanting to frighten people to death" [*Newsday,* 4/14/2000, "City Nixes Malathion Spraying," by Curtis L. Taylor and Dan Morrison].

8. ArboNET currently monitors 14 arboviral diseases, including West Nile, dengue, and SLE. *See:* Lindsey NP, Brown JA, Kightlinger L,

Rosenberg L, Fischer M; ArboNET Evaluation Working Group. "State health department perceived utility of and satisfaction with ArboNET, the U.S. National Arboviral Surveillance System." *Public Health Rep.* 2012 Jul–Aug; 127(4): 383–90.

9. Lee BY, Biggerstaff BJ. "Screening the United States blood supply for West Nile Virus: a question of blood, dollars, and sense." *PLoS Med.* 2006 Feb; 3(2): e99.

10. Roehr B. "Texas records worst outbreak of West Nile virus on record." *BMJ.* 2012 Sep 6; 345: e6019.

11. A retrospective analysis of data collected in New York State in 2000 confirmed that the crow deaths are a reliable marker for early detection of WNV: Eidson M, Kramer L, Stone W, Hagiwara Y, Schmit K; New York State West Nile Virus Avian Surveillance Team. "Dead bird surveillance as an early warning system for West Nile virus." *Emerg Infect Dis.* 2001 Jul–Aug; 7(4): 631–5.

12. Steinhauer J, Miller J. "In New York Outbreak, Glimpse of Gaps in Biological Defenses." *New York Times,* October 11, 1999

13. See: Jones KE, Patel NG, Levy MA, Storeygard A, Balk D, Gittleman JL, Daszak P. "Global trends in emerging infectious diseases." *Nature.* 2008 Feb 21; 451(7181): 990–3; Also: Taylor LH, Latham SM, Woolhouse ME. "Risk factors for human disease emergence." *Philos Trans R Soc Lond B Biol Sci.* 2001 Jul 29; 356(1411): 983–9.

14. In 2007, the American Veterinary Medical Association and the American Medical Association formed a task force that helped launch the One Health Initiative, which promotes an integrated approach to protecting human and animal health in the United States and around the world. See: King LJ, Anderson LR, Blackmore CG, Blackwell MJ, Lautner EA, Marcus LC, Meyer TE, Monath TP, Nave JE, Ohle J, Pappaioanou M, Sobota J, Stokes WS, Davis RM, Glasser JH, Mahr RK. "Executive summary of the AVMA One Health Initiative Task Force report." *J Am Vet Med Assoc.* 2008 Jul 15; 233(2): 259–61.

Chapter 2. The McConnon Strain

Chapter 2 draws on interviews with Patrick McConnon.

1. Six years later, in 1988, Patpong and other red-light districts in Thailand became local epicenters of an outbreak of a previously unknown illness called acquired immunodeficiency syndrome (AIDS). AIDS-inspired efforts to promote safe sex in Patpong are described in: Timm M. "Deadly serious humour for the 'go-go girls'." "Thailand. *AIDS Action.* 1989; Dec: 4. See also Hanenberg R, Rojanapithayakorn W. "Changes in prostitution and the AIDS epidemic in Thailand. "*AIDS Care* 1998; 10: 69–79.

2. *P. vivax* parasites, though rarely causing death, can remain dormant in the liver of humans for months or years, erupting into the bloodstream intermittently to cause episodes of severe illness.

3. Llamzon BS, Gordon RM. *Horyo: Memoir of an American POW.* London: Continuum International Publishing Group; 1999.

4. *P. falciparum* mutations that confer CQ-resistance apparently emerged independently in four places: in the Thailand–Cambodia border region, in two places in South America (Colombia and Venezuela), and in Papua New Guinea. (*See:* Wernsdorfer WH, Payne D. "The dynamics of drug resistance in *Plasmodium falciparum*." *Pharmacol Ther* 1991; 50: 95–121; Ebisawa I, Fukuyama T, Kawamura Y. "Additional foci of chloroquine-resistant falciparum malaria in East Kalimantan and West Irian, Indonesia." *Trop Geogr Med* 1976; 28: 349–54; and Van Dijk WJ. "Mass chemoprophylaxis with chloroquine additional to DDT indoor spraying; report on a pilot project in the Demta area, Netherlands New Guinea." *Trop Geogr Med* 1958; 10: 379–84.)

5. Packard RM. *The Making of a Tropical Disease: A Short History of Malaria.* Baltimore, MD: Johns Hopkins University Press; 2008. *Author's Note*: I contacted Dr. Packard to confirm the location of Pailin. It is located within Cambodia, not far from the Cambodia-Thailand border.

6. Some authorities include a section of Yunnan Province, China, as part of the Golden Triangle. In recent years the Thai tourist industry has begun using the term "Golden Triangle" to denote the meeting point of the Thailand, Laos, and Myanmar borders, at the junction of the Mekong and Ruark Rivers.

7. Information about the spread of drug-resistant malaria strains from Cambodia and Thailand may be found in the following articles:

   • *Sulfadoxine-pyrimethamine (SP):* Roper C, Pearce R, Nair S, Sharp B, Nosten F, Anderson T. "Intercontinental spread of pyrimethamine-resistant malaria." *Science* 2004; 305: 1124.

   • *Chloroquine(CQ):* Ariey F, Fandeur T, Durand R, Randrianariv-elojosia M, Jambou R, Legrand E, et al. "Invasion of Africa by a single PFCRT allele of South East Asian type." *Malar J* 2006; 26(5): 34.

   • *Artesunate-mefloquine:* Wongsrichanalai C, Meshnick SR. "Declining artesunate-mefloquine efficacy against falciparum malaria on the Cambodia–Thailand border." *Emerg Infect Dis* 2008; 14: 716–9. This article describes the detection in Cambodia and Thailand of mutations that confer resistance to newer drugs, such as mefloquine and artemisinin.

Chapter 3. Sorrow and Statistics

Chapter 3 drew on interviews with James W. Buehler and Steve Solomon

1. *Fullfillment. Memoirs of a Criminal Court Judge* by David Vanek. Dundurn Press, Toronto and Oxford, and the Osgoode Society for Canadian Legal History. 1999. Page 290.

2. *Cardiac Arrest, A True Account of Stolen Lives,* by Sarah Spinks. Doubleday Canada Limited, Toronto, 13985. Page 83.

3. Solomon SL, Wallace EM, Ford-Jones EL, Baker WM, Martone WJ, Kopin IJ, Critz AD, Allen JR. "Medication

errors with inhalant epinephrine mimicking an epidemic of neonatal sepsis." *Engl J Med.* 1984 Jan 19; 310(3): 166–70.

4. Transcript of the Royal Commission of Inquiry into Certain Deaths at the Hospital for Sick Children and Related Matters. Transcript of Evidence for January 24, 1984. Volume 91. Pages 418.

5. This saying is variously attributed to Irving J. Selikoff (1915–92), who pioneered occupational medicine, and Sir Austin Bradford Hill (1897–1991), who pioneered the use of random clinical trials.

6. Buehler JW, Smith LF, Wallace EM, Heath CW Jr, Kusiak R, Herndon JL. "Unexplained deaths in a children's hospital. An epidemiologic assessment." *N Engl J Med.* 1985 Jul 25; 313(4): 211–6.

7. Report of the Royal Commission of Inquiry into Certain Deaths at the Hospital for Sick Children and Related Matters. Ministry of the Attorney General, 1984. Page 41.

8. Some doctors who participated in the June 1980 slowdown also went on strike for a week during the fall, from October 30 to November 5. During that week, clinical fellows and staff cardiologists served on the cardiology ward as substitutes for the strikers. *See: Mortality on the Cardiology Service of a Children's Hospital in Toronto, Canada* (the "Atlanta Report"), page 3. This report was submitted to the Canadian Minister of Health on February 16, 1983, by Clark W. Heath, Jr., Lesbia F. Smith, James W. Buehler, and Evelyn M. Wallace. Its conclusions are summarized in Buehler JW, Smith LF, Wallace EM, Heath CW Jr, Kusiak R, Herndon JL. "Unexplained deaths in a children's hospital. An epidemiologic assessment." *N Engl J Med.* 1985 Jul 25; 313(4): 211–6.

9. Of patients with a known room number, the location was the Ward 4A infant room for 22 of 27 babies (81.5%) and the 4B infant room for 3 of 6 (50%). *Mortality on the Cardiology Service of a Children's Hospital in Toronto, Canada* (the "Atlanta Report"), page 15.

10. Berkelman RL, Martin D, Graham DR, Mowry J, Freisem R, Weber JA, Ho JL, Allen JR. "Streptococcal wound infections caused by a vaginal carrier." *JAMA.* 1982 May 21; 247(19): 2680–82.

11. During the Grange Commission, David Hunt, Counsel for the Attorney General and Solicitor General of Ontario (Crown Attorneys and the Coroner's Office) reviewed the sequence and timing of the deaths as part of his cross-examination of Phyllis Trayner. *See: Cardiac Arrest, A True Account of Stolen Lives*, by Sarah Spinks. Doubleday Canada Limited, Toronto, 1985. Page 163.

12. *Death Shift* by Ted Bissland, Methuen & Company, 1984. Chapter 8.

13. Transcript of the Royal Commission of Inquiry into Certain Deaths at the Hospital for Sick Children and Related Matters. Transcript of Evidence for January 24, 1984. Volume 91. Page 577.

14. Transcript of the Royal Commission of Inquiry into Certain Deaths at the Hospital for Sick Children and Related Matters. Transcript of Evidence for January 24, 1984. Volume 91. Pages 427–428.

15. Transcript of the Royal Commission of Inquiry into Certain Deaths at the Hospital for Sick Children and Related Matters. Transcript of Evidence for January 24, 1984. Volume 91. Pages 536–544 (data on access to the cardiology ward by doctors) and pages 556–560 (data on access by nursing supervisors and teaching team leaders). The lawyer for Nurse Trayner continued this line of questioning, especially in regard to nursing supervisors (see: Transcript of Evidence for January 25, 1984. Volume 92. Pages 794–809).

16. Transcript of the Royal Commission of Inquiry into Certain Deaths at the Hospital for Sick Children and Related Matters. Transcript of Evidence for January 25, 1984. Volume 92. Pages 776–779.

17. Transcript of the Royal Commission of Inquiry into Certain Deaths at the Hospital for Sick Children and Related Matters.

Transcript of Evidence for January 26, 1984. Volume 93. Pages 870–71.

18. Transcript of the Royal Commission of Inquiry into Certain Deaths at the Hospital for Sick Children and Related Matters. Transcript of Evidence for January 24, 1984. Volume 91. Page 553.

19. Transcript of the Royal Commission of Inquiry into Certain Deaths at the Hospital for Sick Children and Related Matters. Transcript of Evidence for January 25, 1984. Volume 92. Page 650–1.

20. Report of the Royal Commission of Inquiry into Certain Deaths at the Hospital for Sick Children and Related Matters. Ministry of the Attorney General, 1984. Page 222.

21. Inaccurate digoxin measurements in infants caused by cross-reaction with naturally occurring chemicals have been documented by David Seccombe and his colleagues at the Shaughnessey and Vancouver General Hospitals. (See: Seccombe DW, Pudek MR. "Digoxin-like immunoreactive substances in the perinatal period." *Lancet.* 1987 Apr 25; 1[8539]: 983.) In the *Report of the Royal Commission of Inquiry into Certain Deaths at the Hospital for Sick Children and Related Matters*, Justice Grange acknowledged Secommbe's findings but concluded that "the greatest amount of the [cross-reactive] substance detected in anyone's research to date (4.1 ng/ml in Dr. Seccombe's tests) is miniscule compared with some of the readings . . . encountered in the children whose deaths we are investigating" (Page 26). Justice Grange also stated that, due to difficulties in measuring digoxin levels, "the results of the tests on exhumed tissue were never offered as proof of an overdose of digoxin, only as proof of the presence of digoxin" (Page 31).

22. The theory linking Charles Smith to the autopsies of the infants who died on the cardiology ward at Sick Kids is laid out in: Hamilton G. *The Nurses Are Innocent. The Digoxin Poisoning Fallacy.* Dundurn, Toronto. 2011. Chapter 25. The implications of a 2005–2007 review of 45 cases of

criminally suspicious child deaths involving testimony from Charles Smith are discussed in: Glancey GD, Regehr C. "From Schadenfreude to Contemplation: Lessons for Forensic Experts." *J Am Acad Psychiatry Law* 40: 81–8, 2012.

23. Contemporary accounts include: *Death Shift*, by Ted Bissland (Methuen & Company, 1984) and *Cardiac Arrest, A True Account of Stolen Lives*, by Sarah Spinks (Doubleday Canada Limited, Toronto, Canada, 1985).

24. Hamilton G. *The Nurses Are Innocent. The Digoxin Poisoning Fallacy.* Dundurn, Toronto. 2011

25. Transcript of the Royal Commission of Inquiry into Certain Deaths at the Hospital for Sick Children and Related Matters. Transcript of Evidence for January 24, 1984. Volume 91. Page 499.

26. Yorker BC, Kizer KW, Lampe P, Forrest AR, Lannan JM, Russell DA. "Serial murder by healthcare professionals." *Forensic Sci.* 2006 Nov; 51(6): 1362–71. In a discussion of "Intervention and Prevention" the article notes that hospitals are favorable environments for healthcare serial killer events, due to "easy access to injectable medications, availability of patients with intravenous lines, reduced oversight during evening and night shifts, the frequent use of float nursing personnel, and less than routine quality assurance activities that may increase the likelihood of these crimes going undetected."

27. Yorker BC. "Hospital epidemics of factitious disorder by proxy." *In The spectrum of factitious disorders.* Washington, DC. American Psychiatric Press, 1996.

28. Possible motives of firefighters who set fires are considered in: *The National Volunteer Fire Council Report on the Firefighter Arson Problem. Contexts, Considerations, and Best Practices*, 2011 (http://www.nvfc.org/files/documents/FF_Arson_Report_FINAL.pdf) and *Special Report: Firefighter Arson, U.S. Fire Administration/Technical Report Series, USFA-TR-141/January 2003*, Federal Management Association, U.S. Department of Homeland Security (http://www.usfa.fema.gov/downloads/pdf/publications/tr-141.pdf).

29. "Sick Kids Cold Case Getting Colder." Michele Mandel, *Toronto Sun.* March 5, 2011 (http://www.torontosun.com/news/columnists/michele_mandel/2011/03/05/17507076.html).

Chapter 4. Obession or Inspiration

Chapter 4 drew on interviews with James S. Marks and Joseph McDade

1. Founded as the Communicable Disease Center in 1947, CDC's name was changed in 1970 to the Center for Disease Control and in 1992 to the Centers for Disease Control and Prevention. Its acronym has always been "CDC."
2. "In vitro" refers to tests conducted in test tubes and laboratory dishes; "in vivo" refers to tests conducted in animals.
3. *John Adams*, by David McCullough. Simon & Schuster, New York, NY, 2001. Page 446.
4. Dr. Shepard, who headed the CDC Leprosy and Rickettsia Branch for more than 30 years, died on February 18, 1985. The Charles C. Shepard Science Award—which can be given for scientific publications or for lifetime scientific achievement—was established in his honor in 1986.
5. The Bellevue-Stratford Hotel re-opened in 1979 under a new name.
6. The association between swine flu vaccination and Guillain-Barré syndrome (GBS) was further confirmed by a CDC study published in 1979 (Schonberger LB, Bregman DJ, Sullivan-Bolyai JZ, Keenlyside RA, Ziegler DW, Retailliau HF, Eddins DL, Bryan JA. "Guillain-Barre syndrome following vaccination in the National Influenza Immunization Program, United States, 1976–1977." *Am J Epidemiol.* 1979 Aug; 110(2): 105–23). A reassessment performed in 1991 came to the same conclusion (Safranek TJ, Lawrence DN, Kurland LT, Culver DH, Wiederholt WC, Hayner NS, Osterholm MT, O'Brien P, Hughes JM. "Reassessment of the association between Guillain-Barré

syndrome and receipt of swine influenza vaccine in 1976–1977: results of a two-state study. Expert Neurology Group. *Am J Epidemiol.* 1991 May 1; 133(9): 940–51). The 1979 study found that the period of increased risk was primarily within the first five weeks after vaccination, with some risk remaining for nine to ten weeks." The reassessment study found no increased risk beyond the first six weeks after vaccination.

The reasons for the association between GBS and the 1976 swine flu vaccine remain unknown. Vaccine production methods and safety standards have changed since 1976, resulting in fewer adverse reactions. Today's seasonal flu vaccines have excellent safety profiles (*see*: Vellozzi C et al. "Safety of trivalent inactivated influenza vaccines in adults: background for pandemic influenza vaccine safety monitoring." *Vaccine* 2009; 27: 2114–20). Moreover, no increase in adverse events following vaccination was observed during the 2009–2010 H1N1 pandemic (Centers for Disease Control and Prevention. Safety of Influenza A [H1N1] 2009 Monovalent Vaccines—United States, October 1–November 24, 2009. *Morbidity and Mortality Weekly Report,* December 11, 2009 / 58(48); 1351–56).

7. Thirty-two years later, on April 30, 2009, during the H1N1 influenza pandemic, Dr. Sencer told a CNN reporter that in 1976 health officials "acted on the best knowledge that we had and believed that we were doing the right thing. . . . [But] we know a lot more about viruses than we did then." See also: Sencer DJ, Millar JD. "Reflections on the 1976 swine flu vaccination program." *Emerg Infect Dis.* Volume 12, Number 1—January 2006.

8. Marks JS, Tsai TF, Martone WJ, Baron RC, Kennicott J, Holtzhauer FJ, Baird I, Fay D, Feeley JC, Mallison GF, Fraser DW, Halpin TJ. "Nosocomial Legionnaires' disease in Columbus, Ohio." *Ann Intern Med.* 1979 Apr; 90(4): 565–69. An LD outbreak was also reported in 1977 in Burlington, Vermont (Broome CV, Goings SA, Thacker SB, Vogt RL, Beaty

HN, Fraser DW. "The Vermont epidemic of Legionnaires' disease." Ann Intern Med. 1979 Apr; 90(4): 573–77).

9. An ehrlichial species that causes a human infection called Sennetsu fever, characterized by fever and swollen lymph nodes, was identified in Japan in 1953 and later reported in the Far East and Southeast Asia.

10. Ganguly S, Mukhopadhayay SK. "Tick-borne ehrlichiosis infection in human beings." *J Vector Borne Dis.* 2008 Dec; 45(4): 273–80.

11. A list of pathogens discovered between 1972 and 2004 may be found in: Cohen & Powderly: *Infectious Diseases*, 2nd ed. Volume 1, Section 1, Chapter 4. "Emerging and re-emerging pathogens and diseases." Levitt AM, Khan S, Hughes JM. 2004. Mosby, an imprint of Elsevier. Pathogens discovered since 2004 include: a human bocavirus identified in Sweden that causes acute respiratory illness in children (2005); a transplant-associated arenavirus identified in Australia that killed three recipients of liver or kidney transplants from a single donor (2008); a new tickborne phlebovirus identified in Missouri (the Heartland virus) that causes a severe febrile illness (2012); and a new SARS-like coronavirus identified in the Middle East (2012).

Chapter 5. Dangerous Desserts

Chapter 5 drew on interviews with Richard N. Danila, Craig W. Hedberg, Thomas Hennessy, and Michael T. Osterholm.

1. Scallan E, Hoekstra RM, Angulo FJ, Tauxe RV, Widdowson MA, Roy SL, Jones JL, Griffin PM. Foodborne illness acquired in the United States—major pathogens. Emerg Infect Dis. 2011 Jan; 17(1): 7-15.

2. Hedberg CW, Angulo FJ, White KE, Langkop CW, Schell WL, Stobierski MG, Schuchat A, Besser JM, Dietrich S, Helsel L, Griffin PM, McFarland JW, Osterholm MT.

"Outbreaks of salmonellosis associated with eating uncooked tomatoes: implications for public health." The Investigation Team. *Epidemiol Infect.* 1999 Jun; 122(3): 385–93.

3. Hedberg CW, Korlath JA, D'Aoust JY, White KE, Schell WL, Miller MR, Cameron DN, MacDonald KL, Osterholm MT. "A multistate outbreak of *Salmonella javiana* and *Salmonella oranienburg* infections due to consumption of contaminated cheese." *JAMA.* 1992 Dec 9; 268(22): 3203–07.

4. Hedberg CW, Fishbein DB, Janssen RS, Meyers B, McMillen JM, MacDonald KL, White KE, Huss LJ, Hurwitz ES, Farhie JR, Simmons JL, Braverman LE, Ingbar SH, Schonberger LB, Osterholm MT. "An outbreak of thyrotoxicosis caused by the consumption of bovine thyroid gland in ground beef." *N Engl J Med.* 1987 Apr 16; 316(16): 993–98.

5. Berton Roueché. Annals of Medicine. "A Lean Cusine." *New Yorker*, June 27, 1988, pg. 70–78.

6. *Case Studies in Crisis Communication. Lessons Learned About Protecting America's Food Supply.* Edited by Timothy L. Sellnow and Robert S. Littlefield. Institute for Regional Studies, North Dakota State University, P.O. Box 5075, Fargo ND 58105. 2005. See Chapter 2: "Social Responsibility: Lessons Learned from Schwan's Salmonella Crisis." J.J. McIntyre. (http://www.fooddefense.org/Ncfpd/assets/File/pdf/RC_Lessons_Learned_2005.pdf).

7. Rose Farley. *Tainted Love.* Twin Cities Reader. January 18–24, 1995.

8. Hennessy TW, Hedberg CW, Slutsker L, White KE, Besser-Wiek JM, Moen ME, Feldman J, Coleman WW, Edmonson LM, MacDonald KL, Osterholm MT. "A national outbreak of *Salmonella* enteritidis infections from ice cream." The Investigation Team. *N Engl J Med.* 1996 May 16; 334(20): 1281–86.

9. Mahon BE, Slutsker L, Hutwagner L, Drenzek C, Maloney K, Toomey K, and Griffin PM. "Consequences in Georgia of a nationwide outbreak of *Salmonella* infections: what you don't know might hurt you." *Am J Public Health.* 1999 January; 89(1): 31–35.

10. See: *Food Safety from Farm to Table: a National Food Safety Initiative Report to the President. May 1997.* Food and Drug Administration, U.S. Department of Agriculture, U.S. Environmental Protection Agency, Centers for Disease Control. (http://www.cdc.gov/ncidod/foodsafe/report.htm)

11. *Emerging Infections: Microbial Threats to Health in the United States.* Joshua Lederberg, Robert E. Shope, and Stanley C. Oaks, Jr., eds. Institute of Medicine, National Academies Press, Washington, D.C. 1992.

12. Blaser MJ. "How safe is our food? Lessons from an outbreak of salmonellosis." *N Engl J Med.* 1996 May 16; 334(20): 1324–25. Infection with Salmonella Enteriditis has been has been called a "disease of civilization," and Legionnaires' Disease has been called "a disease of technology" (Chapter 4). In both cases, disease spread is facilitated by modern factors.

13. FoodNet is hosted by the CDC Emerging Infections Program (EIP). Information about FoodNet and EIP may be found at: http://www.cdc.gov/ncezid/dpei/eip/index.html and http://www.cdc.gov/foodnet/

14. Information on PulseNet is available at http://www.cdc.gov/pulsenet/, and information on PulseNet International is available at http://www.pulsenetinternational.org/Pages/default.aspx.

Chapter 6. The Red Mist

Chapter 6 drew on interviews with Ruth Lynfield, Aaron DeVries, Stacy Holzbauer, Richard N. Danila, and James Sejvar.

1. The role of the Spanish interpreter is mentioned in three news reports: Brown, David. "Inhaling pig brains may be cause of new illness." *Washingtonpost.com.* February 4, 2008 (http://www.national-toxic-encephalopathy-foundation.org/PigBrain.pdf); Gajilan, A. "Medical mystery solved in slaughterhouse." *CNNHealth.* February 28, 2008 (http://

articles.cnn.com/2008-02-28/health/medical.mystery_1_
doctors-pig-brain-slaughterhouse?_s=PM:HEALTH);
and Cassels C. "Findings in pork workers with novel neurological illness reported for the first time." *Medscape
Medical News.* April 17, 2008. (http://www.medscape.com/
viewarticle/573193_print).

2. Grady D. "A medical mystery unfolds in Minnesota." *New
York Times.* February 5, 2008.

3. Said G. "Infectious neuropahties." *Neurol Clin* 2007; 25: 115–
37. Infectious causes of polyneuropathy include HIV/AIDS,
Lyme disease, leprosy, herpes zoster, and hepatitis B and C.

4. Kennedy ED, Hall RL, Montgomery SP, Pyburn DG, Jones
JL. "Trichinellosis Surveillance—United States, 2002–2007."
MMWR Surveillance Summary 2009 Dec 4; 58(9): 1–7.

5. Whitfield JT, Pako WH, Collinge J, Alpers MP. "Mortuary
rites of the South Fore and kuru." *Philos Trans R Soc Lond B
Biol Sci.* Nov 27 2008; 363(1510): 3721–24.

6. The index patient is referred to as "Case 1" in: Lachance
DH, Lennon VA, Pittock SJ, Tracy JA, Krecke KN,
Amrami KK, Poeschla EM, Orenstein R, Scheithauer
BW, Sejvar JJ, Holzbauer S, DeVries AS, Dyck PJ. "An
outbreak of neurological autoimmunity with polyradicu-
loneuropathy in workers exposed to aerosolised porcine
neural tissue: a descriptive study." *Lancet Neurol.* 2010
Jan; 9(1): 55–66.

7. Genoways T. "The Spam Factory's Dirty Secret." *Mother
Jones.* July/August 2011 (http://www.motherjones.com/
politics/2011/06/hormel-spam-pig-brains-disease)

8. Brown, David. "Inhaling pig brains may be cause of new illness." *Washingtonpost.com.* February 4, 2008 (http://www.
national-toxic-encephalopathy-foundation.org/PigBrain.
pdf).

9. The history of the Hormel Corporation through the 1980s is
described in: Hage D and Klauda P. *No Retreat, No Surrender:
Labor's War at Hormel.* William Morrow & I Co. 1989.

10. The QPP holding pen was continuously filled with animals who moved along for several hours, giving them sufficient time to acclimate and calm down before they entered the room where they were stunned by electric shock and killed. These procedures—designed in accordance with the ideas of Temple Grandin, the animal expert and autism activist, are both humane and protective of the meat, whose flavor can be affected by high levels of stress-induced lactic acid.

11. Taylor LH, Latham SM, Woolhouse ME. "Risk Factors for human disease emergence." *Philos Trans R Soc Lond B Biol Sci.* 2001 Jul 29; 356(1411): 983–89.

12. Holzbauer SM, DeVries AS, Sejvar JJ, Lees CH, Adjemian J, McQuiston JH, Medus C, Lexau CA, Harris JR, Recuenco SE, Belay ED, Howell JF, Buss BF, Hornig M, Gibbins JD, Brueck SE, Smith KE, Danila RN, Lipkin WI, Lachance DH, Dyck PJ, Lynfield R. "Epidemiologic investigation of immune-mediated polyradiculoneuropathy among abattoir workers exposed to porcine brain." *PLoS One.* 2010 Mar 19; 5(3): e9782 (http://www.ncbi.nlm.nih.gov/pmc/articles/PMC2841649/).

13. The exchange between Lynfield and Wadding is also described in Grady D. "A medical mystery unfolds in Minnesota." *New York Times.* February 5, 2008. (Reference 2.)

14. Gajilan, A. "Medical mystery solved in slaughterhouse." *CNNHealth.* February 28, 2008 (http://articles.cnn.com/2008-02-28/health/medical.mystery_1_doctors-pig-brain-slaughterhouse?_s=PM:HEALTH).

15. Eurosurveillance Editorial Team. "Progressive inflammatory neuropathy (PIN) among swine slaughterhouse workers in Minnesota, United States, 2007–2008." *Euro Surveill.* 2008 Feb 21; 13(8). pii: 8047.

16. Adjemian JZ, Howell J, Holzbauer S, Harris J, Recuenco S, McQuiston J, Chester T, Lynfield R, DeVries A, Belay E, Sejvar. "A clustering of immune-mediated polyradiculoneuropathy among swine abattoir workers exposed to aerosolized

porcine brains, Indiana, United States." *Int J Occup Environ Health.* 2009 Oct–Dec; 15(4): 331–88.

17. Collins TR. "Surveillance Continues on Pork Plant Workers Diagnosed with 'Mystery' Neuropathy." *Neurology Today.* 16 October 2008; Volume 8(20), p 18–20. (http://www.aan.com/elibrary/neurologytoday/?event=home.showArticle&id=ovid.com:/bib/ovftdb/00132985-200810160-00009)

18. Bell H. "Inspector Lachance." *Minnesota Medicine,* November 2008 (http://www.minnesotamedicine.com/Default.aspx?tabid=2712)

19. Cassels C. "Findings in pork workers with novel neurological illness reported for the first time." *Medscape Medical News.* April 17, 2008. (http://www.medscape.com/viewarticle/573193_print).

20. Centers for Disease Control and Prevention (CDC). "Update: outbreak of Nipah virus—Malaysia and Singapore, 1999." *Morb Mortal Wkly Rep.* 1999 Apr 30; 48(16): 335–37.

21. Paton NI, Leo YS, Zaki SR, Auchus AP, Lee KE, Ling AE, Chew SK, Ang B, Rollin PE, Umapathi T, Sng I, Lee CC, Lim E, Ksiazek TG. "Outbreak of Nipah-virus infection among abattoir workers in Singapore." *Lancet.* 1999 Oct 9; 354(9186): 1253–56.

22. Carrieri MP, Tissot-Dupont H, Rey D, Brousse P, Renard H, Obadia Y, Raoult D. "Investigation of a slaughterhouse-related outbreak of Q fever in the French Alps." *Eur J Clin Microbiol Infect Dis.* 2002 Jan; 21(1): 17–21.

23. Marmion BP, Ormsbee RA, Kyrkou M, Wright J, Worswick D, Cameron S, Esterman A, Feery B, Collins W. "Vaccine prophylaxis of abattoir-associated Q fever." *Lancet.* 1984 Dec 22; 2(8417–18): 1411–14.

24. Wilson LE, Couper S, Prempeh H, Young D, Pollock KG, Stewart WC, Browning LM, Donaghy M. "Investigation of a Q fever outbreak in a Scottish co-located slaughterhouse and cutting plant." *Zoonoses Public Health.* 2010 Dec; 57(7–8): 493–8.

25. Dijkstra F, van der Hoek W, Wijers N, Schimmer B, Rietveld A, Wijkmans CJ, Vellema P, Schneeberger PM. "The 2007–2010 Q fever epidemic in The Netherlands: characteristics of notified acute Q fever patients and the association with dairy goat farming." *FEMS Immunol Med Microbiol.* 2012 Feb; 64(1): 3–12.

26. Centers for Disease Control and Prevention (CDC). "Q fever among slaughterhouse workers—California." *MMWR Morb Mortal Wkly Rep.* 1986 Apr 11; 35(14): 223–6.

27. Lachance DH, Lennon VA, Pittock SJ, Tracy JA, Krecke KN, Amrami KK, Poeschla EM, Orenstein R, Scheithauer BW, Sejvar JJ, Holzbauer S, DeVries AS, Dyck PJ. "An outbreak of neurological autoimmunity with polyradiculoneuropathy in workers exposed to aerosolised porcine neural tissue: a descriptive study." *Lancet Neurol.* 2010 Jan; 9(1): 55–66.

28. *The Sign of the Four* (1890) by Sir Arthur Conan Doyle.

29. Gamaleia N. "Etude sur la rage paralytique chez l'homme. (Study of human rabies.)" *Ann Inn Pasteur* 1887; 1: 63–83.

30. Rivers T, Schwenker F. "Encephalomyelitis accompanied by myelin destruction experimentally produced in monkeys." *J Exp Med* 1935; 61: 689–702. In 1955, scientists also induced allergic neuritis (inflammation of the nerves) in laboratory animals through injection of emulsified brain tissue. See: Waksman BH and Adams RD. "Allergic neuritis: an experimental disease of rabbits induced by the injection of peripheral nervous tissue and adjuvants." *J Exp Med.* 1955 Aug 1; 102(2): 213–36.

31. Hemachuda T, Griffin D, Giffels B, Johnson R, Moser A, Phanuphak P," "Myelin basic protein as an encephalitogen in encephalomyelitis and polyneuritis following rabies vaccination." *N Engl J Med* 1987; 316: 369–74.

32. Figueras A, Morales-Olivas FJ, Capella D, Palop V, Laporte JR. "Bovine gangliosides and acute motor polyneuropathy." *BMJ* 1992; 305. The implications of this study are discussed in: Willison HJ, Wraith DC. "A hazardous vapour trail from abattoir to neuropathy clinic." *Lancet Neurol.* 2010 Jan; 9(1): 22–4.

33. Oldstone MB. "Molecular mimicry and immune-mediated diseases." *FASEB J.* 1998 Oct; 12(13): 1255–65.
34. Tracy JA and Dyck PJB. "Auto-immune polyradiculoneuropathy and a novel IgG biomarker in workers exposed to aerosolized porcine brain." *Journal of the Peripheral Nervous System.* Volume 16, Issue s1, June 2011, Pages: 34–37.
35. Cajigal S. "Antibody pattern discovered for pork plant illness." Neuro *Today.* 5 June 2008; Volume 8(11); p3 (http://www.aan.com/elibrary/neurologytoday/?event=home.showArticle&id=ovid.com:/bib/ovftdb/00132985-200806050-00004).
36. Centers for Disease Control and Prevention (CDC). "Investigation of progressive inflammatory neuropathy among swine slaughterhouse workers–Minnesota, 2007–2008." *MMWR Morb Mortal Wkly Rep.* 2008 Feb 8; 57(5): 122–4.
37. DeAngelis TM and Shen L. "Outbreak of progressive inflammatory neuropathy following exposure to aerosolized porcine neural tissue." *Mt Sinai J Med.* 2009 Oct; 76(5): 442–7.
38. Because progressive inflammatory neuropathy (PIN) did not turn out to be a progressive disease, as first believed, the Mayo Clinic neurologists prefer to call it an "occupational autoimmune polyradiculoneuropathy," a term that reflects its discovery in an occupational setting, its autoimmune causation, and its impact on the peripheral nerves and nerve roots. (A "polyradiculoneuropathy" is a disease of the peripheral nerves and the nerve roots.) See: Rukovets O. "Animal Model Mirrors Human Form of Occupational Neuropathy in Pork Plant Workers." *Neurology Today.* 19 January 2012; Volume 12(2); pp 24–27. (Reference 40)
39. Meeusen JW, Klein CJ, Pirko I, Haselkorn KE, Kryzer TJ, Pittock SJ, Lachance DH, Dyck PJ, Lennon VA. "Potassium channel complex autoimmunity induced by inhaled brain tissue aerosol." *Ann Neurol.* 2012 Mar; 71(3): 417–26.
40. Rukovets O. "Animal Model Mirrors Human Form of Occupational Neuropathy in Pork Plant Workers." *Neurology Today.* 19 January 2012; Volume 12(2); pp 24–27

41. Frolov, RV Bagati A, Casino B, Singh S. "Potassium Channels in Drosphila: Historical Breakthroughs, Significance, and Perspectives." *J. Neurogenetics,* 26(3–4): 275–290, 2012; Kamb A, Iverson LE, Tanouye MA. "Molecular characterization of Shaker, a Drosophila gene that encodes a potassium channel." Cell 50, 40513, 1987.

42. The patients' recoveries correlated with decreasing levels (or titers) of neuronal self-antibodies. For most patients, the titers fell below the detection level within three to eighteen months after exposure to the brain-mist ended (see Reference 39). Additional data indicating that the level of anti-VGPC self-antibodies correlates with the level of a patient's symptoms is provided in: Meeusen JW, Lennon VA, Klein CJ. "Immunotherapy-responsive pain in an abattoir worker with fluctuating potassium channel-complex IgG." *Neurology.* 2012 Oct 23; 79(17): 1824–25.

43. Fitzgerald S. "Condition of pork plant workers improving after developing neuropathy." *Neurol Today* 2009; 9: 18–20.

44. Baier E. "Workers sickened at pork plant still wait for compensation." MPRnews. Minnesota Public Radio, March 31, 2010 (http://minnesota.publicradio.org/display/web/2010/03/31/pork-illness-compensation).

Chapter 7. A Normal Spring

Chapter 7 drew on interviews with Jay Butler, Jamie Childs, Thomas Hennessy, James Hughes, Patrick McConnon, and Stuart Nichol. Information about the activities of Bruce Tempest, Mack Sewell, and James Cheek was gathered by the author in 1995 as part of an internal CDC evaluation of the HPS outbreak response.

1. Peters CJ, Olshaker M. *Virus Hunter. Thirty Years of Battling Hot Viruses Around the World.* Anchor Books. Doubleday. 1997. Page 8.

2. Harper DR, Meyer AS. *Of Mice, Men, and Microbes: Hantavirus.* (1999) San Diego: Academic Press.

3. Duchin JS, Koster FT, Peters CJ, Simpson GL, Tempest B, Zaki SR, Ksiazek TG, Rollin PE, Nichol S, Umland ET, et al. "Hantavirus pulmonary syndrome: a clinical description of 17 patients with a newly recognized disease." The Hantavirus Study Group. *N Engl J Med.* 1994 Apr 7; 330(14): 949–55.

4. Grady D. "Death at the Corners." *Discover Magazine.* December 1993 Issue.

5. The phosgene theory is described in: Garrett L. *The Coming Plague. Newly Emerging Diseases in a World Out of Balance.* Farrar, Straus and Giroux, New York, 1994, pgs. 530–31, and in: Peters CJ, Olshaker M. *Virus Hunter. Thirty Years of Battling Hot Viruses Around the World.* Anchor Books. Doubleday, 1997, p.10.

6. A detailed description of the irradiation of the clinical specimens and the antibody testing may be found in: Peters CJ, Olshaker M. *Virus Hunter. Thirty Years of Battling Hot Viruses Around the World.* Anchor Books. Doubleday. 1997. Pgs. 26–27 and 31.

7. Lee PW, Amyx HL, Yanagihara R, Gajdusek DC, Goldgaber D, Gibbs CJ Jr. "Partial characterization of Prospect Hill virus isolated from meadow voles in the United States." *J Infect Dis.* 1985 Oct; 152(4): 826–9.

8. Childs JE, Korch GW, Glass GE, LeDuc JW, Shah KV. "Epizootiology of Hantavirus infections in Baltimore: isolation of a virus from Norway rats, and characteristics of infected rat populations." *Am J Epidemiol.* 1987 Jul; 126(1): 55–68.

9. Childs and his colleagues found that some individuals had antibodies to the Seoul virus in their blood sera, suggesting that they had been exposed but had not been sick (see Reference 8). Decades later, two cases of human disease were attributed to infection with Seoul virus, one in the United

States and one in the United Kingdom. See: Woods Palekar R, Kim P, Blythe D, de Senarclens O, Feldman K, Farnon EC, Rollin PE, Albariño CG, Nichol ST, Smith M. "Domestically acquired Seoul virus causing hemorrhagic fever with renal syndrome-Maryland, 2008." *Clin Infect Dis.* 2009 Nov 15; 49(10): e109–12, and Jameson LJ, Logue CH, Atkinson B, Baker N, Galbraith SE, Carroll MW, Brooks T, Hewson R. "The continued emergence of hantaviruses: isolation of a Seoul virus implicated in human disease, United Kingdom, October 2012." *Euro Surveill.* 2013 Jan 3; 18(1): 4–7.

10. Rodriguez LL, Letchworth GJ, Spiropoulou CF, Nichol ST. "Rapid detection of vesicular stomatitis virus New Jersey serotype in clinical samples by using polymerase chain reaction." *J Clin Microbiol.* 1993 Aug; 31(8): 2016–20.

11. Nichol ST, Spiropoulou CF, Morzunov S, Rollin PE, Ksiazek TG, Feldmann H, Sanchez A, Childs J, Zaki S, Peters CJ. "Genetic identification of a hantavirus associated with an outbreak of acute respiratory illness." *Science.* 1993 Nov 5; 262(5135): 914–7.

12. Zaki SR, Greer PW, Coffield LM, Nolte KB, Zumwalt R, Umiand ET, Feddersen RM, Foucar K, Ruo SL, Rollin P, Ksiazek T, Nichol S, Peters CJ. "Outbreak of hantavirus-associated illness in the United States: immunohistochemical localization of viral nucleoproteins to endothelial cells in human tissues." *Lab Invest* 1994, 70: 129A (Abstract)

13. Zaki SR, Greer PW, Coffield LM, Goldsmith CS, Nolte KB, Foucar K, Feddersen RM, Zumwalt RE, Miller GL, Khan AS. "Hantavirus pulmonary syndrome. Pathogenesis of an emerging infectious disease." *Am J Pathol.* 1995 Mar; 146(3): 552–79.

14. Saltzstein. "Southwest's 'Navajo Flu' Deadly But Not Navajo." *American Journalism Review,* October 1993.

15. Leovy J, Cheevers J. *Los Angeles Times.* June 2, 1993. "Visiting Navajo Children Barred From L.A. School : Health: Officials feared students would be exposed to deadly illness. Medical authorities say they overreacted."

16. Zeitz PS, Butler JC, Cheek JE, Samuel MC, Childs JE, Shands LA, Turner RE, Voorhees RE, Sarisky J, Rollin PE, et al. "A case-control study of hantavirus pulmonary syndrome during an outbreak in the southwestern United States." *J Infect Dis.* 1995 Apr; 171(4): 864–70

17. Childs JE, Krebs JW, Ksiazek TG, Maupin GO, Gage KL, Rollin PE, Zeitz PS, Sarisky J, Enscore RE, Butler JC, et al. "A household-based, case-control study of environmental factors associated with hantavirus pulmonary syndrome in the southwestern United States." *Am J Trop Med Hyg.* 1995 May; 52(5): 393–7.

18. Childs JE, Ksiazek TG, Spiropoulou CF, Krebs JW, Morzunov S, Maupin GO, Gage KL, Rollin PE, Sarisky J, Enscore RE, et al. "Serologic and genetic identification of *Peromyscus maniculatus* as the primary rodent reservoir for a new hantavirus in the southwestern United States." *J Infect Dis.* 1994 Jun; 169(6): 1271–80.

19. Childs JE, Kaufmann AF, Peters CJ, Ehrenberg RL. "Hantavirus infection–southwestern United States: interim recommendations for risk reduction." Centers for Disease Control and Prevention. *MMWR Recomm Rep.* 1993 Jul 30; 42(RR-11): 1–13.

20. The advice and concerns of the Navajo medicine men are also described in: Peters CJ, Olshaker M. *Virus Hunter. Thirty Years of Battling Hot Viruses Around the World.* Anchor Books. Doubleday. 1997. Pgs. 24–25.

21. Parmenter RR, Brunt JW, Moore DI, Ernest S. (1993). "The hantavirus epidemic in the Southwest: Rodent population dynamics and the implications for transmission of hantavirus-associated adult respiratory distress syndrome (HARDS) in the Four Corners region (Univ. of New Mexico, Albuquerque)," Sevilleta Long-Term Ecological Research Site Pub. No 41.

22. Yates TL, Mills JN, Parmenter CA, Ksaizek TG, Parmenter RR, Vande Castle JR, Calisher CH, Nichol ST, Abbott KD, Young JC, Morrison ML, Beaty BJ, Dunnum JL, Baker RJ,

Salazar-Bravo J, Peters CJ. "The Ecology and Evolutionary History of an Emergent Disease: Hantavirus Pulmonary Syndrome." *BioScience* 52(11): 989–98. 2002.

23. McCormick JB, King IJ, Webb PA, Scribner CL, Craven RB, Johnson KM, Elliott LH, Belmont-Williams R. "Lassa fever. Effective therapy with ribavirin." *N Engl J Med.* 1986 Jan 2; 314(1): 20–6.

24. Huggins JW, Hsiang CM, Cosgriff TM, Guang MY, Smith JI, Wu ZO, LeDuc JW, Zheng ZM, Meegan JM, Wang QN, et al. "Prospective, double-blind, concurrent, placebo-controlled clinical trial of intravenous ribavirin therapy of hemorrhagic fever with renal syndrome." *J Infect Dis.* 1991 Dec; 164(6): 1119–27.

25. The initial ribavirin trial was inconclusive, and the drug was not shown to be effective against HPS in later studies. *See*: Chapman LE, Mertz GJ, Peters CJ, Jolson HM, Khan AS, Ksiazek TG, Koster FT, Baum KF, Rollin PE, Pavia AT, Holman RC, Christenson JC, Rubin PJ, Behrman RE, Bell LJ, Simpson GL, Sadek RF. "Intravenous ribavirin for hantavirus pulmonary syndrome: safety and tolerance during 1 year of open-label experience." Ribavirin Study Group. *Antivir Ther.* 1999; 4(4): 211–9, and Mertz GJ, Miedzinski L, Goade D, Pavia AT, Hjelle B, Hansbarger CO, Levy H, Koster FT, Baum K, Lindemulder A, Wang W, Riser L, Fernandez H, Whitley RJ; Collaborative Antiviral Study Group. "Placebo-controlled, double-blind trial of intravenous ribavirin for the treatment of hantavirus cardiopulmonary syndrome in North America." *Clin Infect Dis.* 2004 Nov 1; 39(9): 1307–13.

26. Crowley MR, Katz RW, Kessler R, Simpson SQ, Levy H, Hallin GW, Cappon J, Krahling JB, Wernly J. "Successful treatment of adults with severe Hantavirus pulmonary syndrome with extracorporeal membrane oxygenation." *Crit Care Med.* 1998 Feb; 26(2): 409–14.

27. Dietl CA, Wernly JA, Pett SB, Yassin SF, Sterling JP, Dragan R, Milligan K, Crowley MR. "Extracorporeal membrane

oxygenation support improves survival of patients with severe Hantavirus cardiopulmonary syndrome." *J Thorac Cardiovasc Surg.* 2008 Mar; 135(3): 579–84.

28. Christian MD, Poutanen SM, Loutfy MR, Muller MP, Low DE. "Severe acute respiratory syndrome." *Clin Infect Dis.* 2004 May 15; 38(10): 1420–27.

29. Zaki SR, Khan AS, Goodman RA, Armstrong LR, Greer PW, Coffield LM, Ksiazek TG, Rollin PE, Peters CJ, Khabbaz RF. "Retrospective diagnosis of hantavirus pulmonary syndrome, 1978-1993: implications for emerging infectious diseases." *Arch Pathol Lab Med.* 1996 Feb; 120(2): 134–9.

30. As of December 31, 1993, 53 cases of HPS had been detected in 14 states. *See:* CDC. "Hantavirus pulmonary syndrome—United States, 1993." *MMWR.* 1994 Jan 28; 43(3): 45–8. In 1994, additional cases were reported in Florida, Virginia, and the northeastern United States. *See:* CDC. "Newly identified hantavirus—Florida, 1994." 1994 Feb 18; 43(6): 99, 105; CDC. "Hantavirus pulmonary syndrome–Virginia, 1993." *MMWR.* 1994 Dec 2; 43(47): 876–7; and CDC. "Hantavirus pulmonary syndrome—Northeastern United States, 1994." *MMWR.* 1994 Aug 5; 43(30): 548-9, 555–66.

31. Centers for Disease Control and Prevention (CDC). "Hantavirus pulmonary syndrome in visitors to a national park—Yosemite Valley, California, 2012." *MMWR* 2012 Nov 23; 61(46): 952.

32. In 1997, fifteen years before the 2012 HPS outbreak, Childs, Peters, Ksiazek, and other CDC colleagues conducted a study in Yosemite and two other national parks (Sequoia/Kings Canyon National Parks in California and Shenandoah National Park in Virginia) to identify ways to keep mice from entering homes and cabins in rural areas during the fall and winter. They found that even minor repairs help prevent rodents from entering homes. (*See:* Mills JN, Johnson JM, Ksiazek TG, Ellis BA, Rollin PE, Yates TL, Mann MO,

Johnson MR, Campbell ML, Miyashiro J, Patrick M, Zyzak M, Lavender D, Novak MG, Schmidt K, Peters CJ, Childs JE. "A survey of hantavirus antibody in small-mammal populations in selected United States National Parks." *Am J Trop Med Hyg*. 1998 Apr; 58[4]: 525–32.) Also in 1997, they conducted a survey to evaluate hantavirus activity among small mammals in 39 national parks in the eastern and central United States. Antibody reactive to the HPS virus was found in deer mice, white-footed mice, rice rats, cotton rats, western harvest mice, three species of voles, and one species of chipmunk. Antibody prevalence among deer mice was highest in the northeast. (*See:* Mills JN, Johnson JM, Ksiazek TG, Ellis BA, Rollin PE, Yates TL, Mann MO, Johnson MR, Campbell ML, Miyashiro J, Patrick M, Zyzak M, Lavender D, Novak MG, Schmidt K, Peters CJ, Childs JE. "A survey of hantavirus antibody in small-mammal populations in selected United States National Parks." *Am J Trop Med Hyg*. 1998 Apr; 58[4]: 525–32.)

33. Renewed U.S. concern about infectious diseases was spurred by the emergence of AIDS and the re-emergence of TB, which demonstrated that viruses and bacteria had not been "conquered" by modern medicine. *See:* Institute of Medicine. *Emerging Infections: Microbial Threats to Health in the United States*. National Academies Press. Washington, DC. 1992.

34. The CDC Special Pathogens Laboratory isolated Sin Nombre virus (SNV) from a deer mouse trapped near the New Mexico home of an HPS patient. Shortly afterwards and independently, USAMRIID isolated SNV from an HPS patient in New Mexico and a mouse trapped in California. *See:* Marshall E, Stone R. "Race to grow hantavirus ends in tie." *Science*. 1993 Dec 3; 262(5139): 1509; and Elliott LH, Ksiazek TG, Rollin PE, Spiropoulou CF, Morzunov S, Monroe M, Goldsmith CS, Humphrey CD, Zaki SR, Krebs JW, Maupin G, Gage K, Childs JE, Nichol ST, Peters CJ. "Isolation of the causative agent of hantavirus pulmonary syndrome." *Am J Trop Med Hyg* 1994, 51: 102–8.

Epilogue. June 2013

1. In late 2013, Chikungunya virus was found for the first time in the Americas, on islands in the Caribbean. Since then, it has spread to forty-four countries and territories throughout the Americas, including the United States, where cases of Chikungunya virus disease have been reported in Florida, Puerto Rico, and the U.S. Virgin Islands.

Acknowledgments

Above all, I would like to thank the medical detectives and scientists who shared their outbreak stories with me. They include James W. Buehler, Jay Butler, Jamie Childs, Richard N. Danila, Jeffrey P. Davis, Aaron DeVries, Annie Fine, Duane J. Gubler, Craig W. Hedberg, Thomas Hennessy, Stacy Holzbauer, James M. Hughes, James J. Kazmierczak, Marci Layton, Ruth Lynfield, James S. Marks, Patrick J. McConnon, Joseph McDade, Stuart T. Nichol, Michael T. Osterholm, James J. Sejvar, and Steven Solomon. Their willingness to talk about their work and their professional and personal experiences made this book possible.

I would like to thank Patrick McConnon, who initiated this project and helped me every step of the way. I also want to thank James M. Hughes for sparking my interest in outbreak stories; Karen Foster for reviewing my use of scientific language and terminology; and Ann Schaetzel for helping me track down a long-sought article. Special thanks also to Annie Fine, Craig W. Hedberg, Thomas Hennessy, James M. Hughes, Rima Khabbaz, Marci Layton, Ruth Lynfield, Ann Schaetzel, and Roland Sutter for comments and advice. Thanks also to Lakesha Robinson and Jennifer Lemmings for research support; to Ed Chow, Lauren Rosenberg, and Jennifer Lemmings for help with graphics and illustrations; and to Jamie Childs, Richard Danila, Kenneth Gage, Craig W. Hedberg, Ruth Lynfield, Craig Manning, Kate McConnon, Stuart Nichol, and Christina Spiropoulou for photographs.

I am grateful to my agent, Joelle Delbourgo, for her advice, persistence and guidance, and my editor, Kristin Kulsavage, for her skill and encouragement.

Finally, I want to thank my colleagues at CDC (too many to name here), as well as family members and friends who saw me through the long process of planning and completing a book. Among others, those family members and friends include: Sean Fischer-Werner, Ryan Fischer-Werner, Megan Fischer-Werner, Lily Fischer-Levitt, Emma Beckerle, Scott Fischer, Florence Levitt, Kate McConnon, Ann Schaetzel, and Fran Schumer.